Maximilian Röhe

Dynamic Model-based Analysis of Oxygen Reduction Reaction in Gas Diffusion Electrodes

Schriftenreihe
des Instituts für Angewandte Materialien
Band 110

Karlsruher Institut für Technologie (KIT)
Institut für Angewandte Materialien (IAM)

Eine Übersicht aller bisher in dieser Schriftenreihe erschienenen Bände finden Sie am Ende des Buches.

Dynamic Model-based Analysis of Oxygen Reduction Reaction in Gas Diffusion Electrodes

by
Maximilian Röhe

KIT Scientific Publishing

Karlsruher Institut für Technologie
Institut für Angewandte Materialien

Dynamic Model-based Analysis of Oxygen Reduction Reaction
in Gas Diffusion Electrodes

Zur Erlangung des akademischen Grades eines Doktor-Ingenieurs
von der KIT-Fakultät für Elektrotechnik und Informationstechnik des
Karlsruher Instituts für Technologie (KIT) genehmigte Dissertation

von Maximilian Röhe, M.Sc.

Tag der mündlichen Prüfung: 28. Januar 2022
Hauptreferentin: Prof. Dr.-Ing. Ulrike Krewer
Korreferent: Prof. Dr.-Ing. Thomas Turek

Impressum

KIT Scientific Publishing

Karlsruher Institut für Technologie (KIT)
KIT Scientific Publishing
Straße am Forum 2
D-76131 Karlsruhe

www.ksp.kit.edu

Print on Demand 2023 – Gedruckt auf FSC-zertifiziertem Papier

ISSN 2192-9963
ISBN 978-3-7315-1234-9
DOI 10.5445/KSP/1000150805

Kurzfassung

Weltweit werden ca. 1 % der gesamten elektrischen Energie für die Chlor-Alkali Elektrolyse aufgewendet. Die für die Elektrolyse benötigte elektrische Energie kann um ca. ein Drittel reduziert werden, indem auf Silberkatalysator basiertende Gasdiffusionselektroden – sog. Sauerstoffverzehrkathoden (englisch: oxygen depolarized cathodes, ODC) – in den Prozess integriert werden. Ungeachtet ihres großen Energieeinsparpotentials sind die Prozesse und Limitierungen innerhalb der ODC weder verstanden noch quantifiziert. Die Einflüsse der Elektrodenstruktur auf die Leistungsfähigkeit und die Dynamik der ODC sind kaum ergründet. Zur Beantwortung der genannten Forschungsfragen wird das erste Modell einer ODC vorgestellt, das das Phasengleichgewicht der Gas-Flüssigkeitsgrenzfläche und strukturbedingte Inhomogenitäten in der Elektrolytverteilung berücksichtigt.

Es wird gezeigt, dass die Stromdichte durch die Verfügbarkeit an gelöstem Sauerstoff im flüssigen Elektrolyten limitiert wird. Bereits im stromlosen Zustand ist nur wenig Sauerstoff in dem Elektrolyten lösbar, der Grund liegt in der hohen Ionenkonzentration und der damit verbundenen niedrigen Wasseraktivität. Durch die Sauerstoffreduktion wird mit zunehmender Stromstärke Wasser verbraucht und Hydroxidionen produziert, die sich nahe der Gas-Flüssigkeitsgrenze anreichern. Dies führt zu einer weiteren Verringerung der Wasseraktivität und somit zu einer weiteren Verschlechterung der Sauerstofflöslichkeit und schließlich zu einer Sauerstoffverarmung.

Auf Grundlage dieser Erkenntnisse wird im zweiten Teil der Einfluss des Wassermassentransports und der Wasseraktivität detaillierter untersucht. Dazu werden zwei vergleichbare ODCs in Messaufbauten mit unterschiedlichen Stofftransportbedingungen in der Flüssigphase mithilfe des dynamischen Dreiphasenmodells analysiert. Es wird gezeigt, dass ein konvektiver Elektrolytstrom nicht nur zu einer höheren Elektrodenperformance, sondern auch zu einer weitaus schnelleren Dynamik führt. Beide Effekte sind auf kürzere Diffusionsschichten in der Flüssigphase und somit auf einen schnelleren Wasser- und Ionenmassentransport zurückzuführen. Die Wasseraktivität beeinflusst die Sauerstoffreduktion nur geringfügig direkt, aber signifikant indirekt über die Sauerstofflöslichkeit.

Im dritten Teil wird der Einfluss der ODC-Struktur auf die Performance und Dynamik untersucht. Ein modelgestützter Vergleich von ODCs mit unterschiedlichen Binder-Katalysator-Verhältnissen mittels elektrochemischer Impedanzspektroskopie führt zu der Erkenntnis, dass Unterschiede in der Performance im Wesentlichen auf die unterschiedlich ausgeprägte Gas-Flüssigkeitsgrenzschicht zurück zu führen sind, die die Menge an gelöstem Sauerstoff signifikant beeinflusst. Die Impedanzspektren aller untersuchten Elektroden legen den Schluss nahe, dass sich die Dreiphasengrenzfläche inhomogen über die gesamte Elektrodentiefe erstreckt, jedoch scheinen die stärker gefluteten Elektrodenteile elektrochemisch nahezu inaktiv zu sein.

Zusammenfassend wird mithilfe der makroskopischen dynamischen Modellierung aufgezeigt, dass zwei Faktoren besonders entscheidend für eine hohe ODC-Performance sind: Eine große Gas-Flüssigkeitsgrenzschicht, die besonders durch die ODC-Struktur beeinflusst werden kann, und ein schneller Wasser- und Ionenmassentransport in der Flüssigphase, der sowohl durch eine konvektive Strömung des Elektrolyten als auch durch die ODC-Struktur begünstigt werden kann. Auf der methodischen Ebene wird demonstriert, dass die dynamische Modellierung ein geeignetes Tool ist, um komplexe Dreiphasensysteme wie das der ODC zu analysieren. Dabei ist besonders die dynamische Ermittlung von Zeitkonstanten zur Separation der sich überlagernden Prozesse unabdingbar.

Abstract

Worldwide approximately 1 % of the total electrical energy is spent on chlor-alkali electrolysis. The demand of electrical energy for the electrolysis can be reduced by about one third by integrating silver catalyst-based gas diffusion electrodes – so-called oxygen depolarized cathodes (ODC) – into the process. Regardless of its large energy saving potential, processes and limitations within ODCs are neither understood nor quantified. Also, the influences of the electrode structure on performance and dynamics are poorly investigated. To address the mentioned research questions, the first model of an ODC is presented that considers the phase equilibrium of the gas-liquid interface and structure-related inhomogeneities in electrolyte distribution.

It is shown that the current density is limited by the availability of dissolved oxygen in the liquid electrolyte. Even at open-cell state, oxygen is poorly soluble in the liquid electrolyte due to the high ion concentration and low water activity. With increasing current density, the oxygen reduction reaction consumes water and produces hydroxide ions which accumulate near the gas-liquid interface. This leads to a further reduction in water activity and thus to a deterioration in oxygen solubility and finally to oxygen depletion.

Next, the influences of water mass transport and water activity are investigated in detail. For this purpose, ODC measurements of two experimental setups differing in mass transport conditions in the liquid phase are analyzed with the dynamic three-phase model. It is demonstrated that a convective electrolyte flow does not only lead to a higher electrode performance, but also to far faster dynamics. Both effects are attributed to shorter mass transport length in the liquid phase and thus to faster water and ion mass transport.

In the third part of this dissertation, the influence of ODC structure on performance and dynamics is examined. By means of electrochemical impedance spectroscopy it is demonstrated that the ODC performance is mainly governed by sizes of the gas-liquid interface, which significantly affects the amount of dissolved oxygen. The model-based analysis of all electrodes suggest that the three-phase interface extends inhomogeneously over the entire electrode depth, but the three-phase area more close to the gas side of the ODC is electrochemically almost inactive.

In conclusion, macroscopic dynamic modeling reveals that two factors are particularly critical for a high ODC performance: A large gas-liquid interface, which can be mainly influenced by the ODC structure, and a fast water and ion mass transport in the liquid phase, which can be favored by a convective flow of the electrolyte and also by the ODC structure. Methodologically it is demonstrated that dynamic modeling is a suitable tool to analyze complex three-phase systems such as ODCs. Especially the dynamic determination of time constants for the separation of the overlapping processes is indispensable.

Acknowledgments

This dissertation is the result of years of research, which were filled by success and joy, but also by hard work and temporary stagnation. At this point I would like to thank all the people, who helped me to finish this project.

First of all, I would like to thank my supervisor Prof. Dr.-Ing. Ulrike Krewer for her guidance, the many helpful advices and for the time and energy she invested in my scientific education. Many thanks go to Prof. Dr.-Ing. Thomas Turek, for being my second referee. Thanks also go to Prof. Dr.-Ing. Marc Hiller for chairing my examination board and to Prof. Dr. Sebastian Kempf and Prof. Dr.-Ing. Sören Hohmann for being part of my examination board.
I would also like to thank my colleagues at the InES for all the inspiring scientific and non-scientific discussions, the great working atmosphere and the joint coffee breaks. My special thanks go to, Fabian Kubannek, Oke Schmidt, Vincent Laue, Florian Baakes, Walter Cistjakov, Marco Heinrich, Wilfried Jansen, Horst Müller, Jasmin Müller, Fridolin Röder, Ina Schunke and Nicolas Wolff.
I would also like to thank the members of my research group for the great cooperation, especially my co-authors Alexander Botz, Barbara Ellendorff, David Franzen, Denis Öhl, Prof. Dr. Wolfgang Schuhmann and Prof. Dr.-Ing. Thomas Turek.
Finally, my thanks go to my family and friends who have supported me during the entire time. Very special thanks go to my parents' neighbor Karsten.

Braunschweig, im Monat Januar 2022 *Maximilian Röhe*

Contents

Kurzfassung i

Abstract iii

Acknowledgments v

1 Introduction 1
1.1 Motivation and purpose 1
1.2 Scope and structure 3

2 Advanced Chlor-Alkali Electrolysis and Oxygen Depolarized Cathodes 5
2.1 Processes of chlor-alkali electrolysis 5
2.1.1 Conventional methods 6
2.1.2 Membrane process with oxygen depolarized cathodes 8
2.2 Processes in oxygen depolarized cathodes 9
2.3 Concluding remarks 11

3 Dynamic Modeling and Analyzing of Oxygen Depolarized Cathodes 13
3.1 Mechanistic and literature models of oxygen depolarized cathodes 13
3.1.1 Mechanistic models of gas diffusion electrodes 14
3.1.2 Models of oxygen depolarized cathodes for chlor-alkali electrolysis 15
3.2 Phase transitions 16
3.3 Reaction kinetics in oxygen depolarized cathodes 18
3.4 Mass transport in oxygen depolarized cathodes 20
3.4.1 Maxwell-Stefan diffusion 21
3.4.2 Fick's diffusion and Stefan-flow 25
3.5 Dynamic analysis of oxygen depolarized cathodes 27
3.6 Concluding remarks 33

4 Processes and Their Limitations in Oxygen Depolarized Cathodes 35
4.1 Introduction 35

4.2 Basic model: Model concept and assumptions 36
4.3 Model equations 39
4.3.1 Electrochemical reaction kinetics, double layer capacitance and internal resistance 39
4.3.2 Mass transport in the gas phase and processes at the gas-liquid interface 40
4.3.3 Mass transport in the thin-film 42
4.3.4 Mass transport in the flooded agglomerates 42
4.3.5 Mass transport in the liquid diffusion layer 43
4.3.6 Boundary conditions 44
4.4 Parameterization 44
4.5 Results and discussion 46
4.5.1 Steady state analysis 46
4.5.2 Dynamic analysis 49
4.5.3 Parameter sensitivities 53
4.6 Conclusions 55

5 In-depth Analysis of Water Mass Transport and Water Activity 57
5.1 Introduction 57
5.2 Experimental 58
5.3 Modelling and parameterization 59
5.3.1 Expanded liquid mass transport model 59
5.3.2 Parameterization and parameter evaluation 62
5.4 Steady state analysis of oxygen depolarized cathodes performance 63
5.5 Dynamic analysis 68
5.6 Conclusions 73

6 In-depth Analysis of Electrolyte Distribution 75
6.1 Introduction 75
6.2 Experimental 77
6.3 Analysis of measurement results 77
6.4 Modelling 81
6.4.1 The domain sub-model 82
6.4.2 The inhomogeneous ODC model 86
6.4.3 Properties and parameters of oxygen depolarized cathodes and representative electrode domains 87
6.5 Comparative model-based analysis of inhomogeneous oxygen depolarized cathodes 91
6.6 Model-based scenario analysis of impedance spectra 99

6.7 Conclusions 103

7 Conclusions **105**

7.1 Summary and conclusions 105

7.2 Open questions and perspective 107

Bibliography **109**

List of Figures **119**

List of Tables **123**

List of Symbols and Abbreviations **125**

A Maxwell-Stefan Diffusion **129**

B Specific Values **133**

C Experimental of Chapter 5 . . . . **137**

C.1 ODC preparation 137

C.2 Experimental set-up with electrolyte convection 137

C.3 Experimental set-up with SECM and stagnant electrolyte 138

C.4 SECM electrode fabrication and measuring 139

D Experimental and Scenario Parameters of Chapter 6 . . . . **141**

D.1 Experimental 141

D.1.1 Electrode preparation 141

D.1.2 Electrochemical characterization 141

D.2 EIS at different current densities 142

D.3 Model parameters of the scenario analysis 143

1 Introduction[1]

1.1 Motivation and purpose

One of the greatest challenges for current and future generations is to mitigate climate change by significantly reducing greenhouse gas emissions. Two key levers for facing this challenge are the climate-friendly provision of energy and the reduction of energy consumption. This thesis focuses on the latter aspect: A significant share of global electricity consumption is accounted for chlorine production. As a basic chemical for 60 % of all chemical products [3], the world wide demand of chlorine has been increased from 42 million tons in 2001 [4] to 52 million tons in 2003 [5] and 89 million tons in 2017 [3]. Using the traditional NaCl-membrane electrolysis, which is the most energy-efficient of all established chlorine production processes [6], the chlorine production in 2017 corresponded to a consumption of 195 800 GWh electric energy – which equals about 1 % of the global consumption of electrical energy [7] – and the emission of 82 million tons of CO_2 [3]. However, further energy optimizations of the traditional NaCl-membrane electrolysis are not attainable, since the savings potential of this process has been practically exhausted [4].

An thermodynamically driven increase of energy efficiency is enabled by a fundamental redesign of the membrane-process: The chlor-alkali electrolysis using oxygen depolarized cathodes (ODC). Here, in contrast to the standard membrane-process, no hydrogen is produced as co-product, but oxygen is consumed at the cathode [8]. The conversion of the process reduces the required electrical energy by about 30 % [6]. The ODC, as the centerpiece of this technique, is a silver catalyst-based gas diffusion electrode (GDE). In ODCs, oxygen is provided in the gas phase, which dissolves into the 30 wt−% NaOH electrolyte, where it reacts at the silver surface with water under the acceptance of electrons to hydroxide ions [5]:

$$O_2 + 2\,H_2O + 4\,e^- \xrightarrow{Ag} 4\,OH^- \qquad (1.1)$$

1 Parts of this chapter have been published in Röhe et al., ChemSusChem, 2019 [1] and Röhe et al., ChemElectroChem, 2019 [2].

To enable both, the liquid and the gas phase within the ODC, it consists in addition to hydrophilic silver particles also out of hydrophobic polytetrafluoroethylene (PTFE). This structure leads to a complex three-phase system, in which many processes interact (cf. fig. 1.1): Mass transport in the gas phase, mass transport in the liquid phase, processes at the gas-liquid interface such as water evaporation of the oxygen solution and the actual oxygen reduction reaction (ORR).

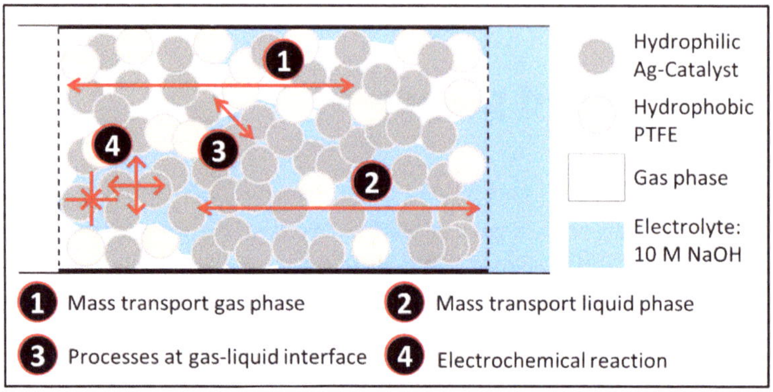

Figure 1.1: Schematic view on processes in oxygen depolarized cathodes.

Irrespective of the great importance of ODCs, the processes and limitations are not well understood, especially not quantitatively. Essential properties like the local distribution of the gaseous and liquid phase within the electrode are unknown [9]. This missing insights harm a knowledge-driven improvement of ODCs.

One approach to analyze the processes separately as well as their mutual interaction and their contribution to the electrode performance is through dynamic modeling.

This approach has already been used successfully in many studies: Dynamic analysis was used for discriminating between reaction kinetic models e.g. for oxygen reduction [10] or methanol oxidation [11], for quantifying the impact of reactant transport, e.g. of oxygen through liquid electrolyte on ORR performance [12], for state estimation [13] or concentration sensing [14].

The objective of this thesis is derived from the open research questions of ODCs and the opportunities of dynamic modeling: *The goal of this dissertation is to gain deeper insights into the processes in ODCs for advanced chlor-alkali electrolysis in order to identify limitations and to provide the fundamentals for knowledge-based electrode improvements.* To reach the goal, the suitable approach of dynamic modeling is used to identify and separate overlapping processes and limitations in ODCs.

1.2 Scope and structure

This dissertation documents the development and analysis of dynamic three-phase models of ODCs. Initially, an overview of ODCs and their processes is given based on the current literature. Subsequently, ODC models are presented, parameterized, analyzed and extended on the basis of intermediate results. This dissertation is structured as follows:
The target of the **second chapter** is to reveal and describe the processes as well as the process conditions of ODCs. After a brief overview of conventional chlor-alkali electrolysis processes, the chlor-alkali electrolysis with ODC and the ODC itself are discussed in detail. At the end of the chapter, the characteristic processes and properties of ODCs and therefore the essentials for ODC model development are summarized.
In the **third chapter**, the processes and system properties previously classified as relevant are considered more in detail. The aim of this chapter is to lay the base for the actual model design. In detail, approaches to model the gas-liquid interface, the processes at the gas-liquid interface, the reaction kinetics and the mass transport are presented. In addition, suitable dynamic methods for the analysis of ODCs are discussed. The findings of these chapters are summarized in a concretization of the research question and in a list of concrete model requirements.
The **fourth chapter** includes the actually model design of the dynamic one-dimensional three-phase model, the validation as well as a model analysis. The model is parameterized using stationary literature data. A subsequent dynamic analysis shows that the dynamic behavior of the ODC is dominated by the water and ion transport in the liquid phase. This leads to the conclusion that in the common understanding of the ODC the influence of the water activity is underrepresented. The results are supported by a sensitivity analysis, which shows that especially parameters influencing the water and ion mass transport in the liquid phase have a high influence on the electrode performance.
In the **fifth chapter** the role of water mass transport and water activity is analyzed in-depth. For this purpose the model is expanded to more accurately represent the influences of mass transport in the liquid phase. Two different measurement setups with different mass transport conditions in the liquid phase[2] were investigated with the dynamic three-phase model by steady-state and dynamic data. It will be shown that a poor mass transport in the liquid phase leads not only to considerably slower dynamics, but also to a much lower performance of the ODC. It is demonstrated that water mass transport and water activity do not significantly influence the ORR directly, but indirectly via the water activity-dependent solubility of oxygen.
Chapter six focuses on the structure of the ODC. This also includes further extensions of the

[2] Measurements performed by Institute of Chemical and Electrochemical Process Engineering, Technische Universität Clausthal, Germany and Institute of Analytical Chemistry — Center for Electrochemical Sciences, Ruhr University Bochum, Germany, published in [2].

model regarding inhomogeneities in electrolyte distribution. The influence of different PTFE-silver ratios to electrode performance and dynamics is evaluated by means of electrochemical impedance spectroscopy.[3] It is demonstrated that the three-phase interface extends inhomogeneously over the entire electrode depth. The simulation results indicate that the more heavily flooded electrode parts are almost electrochemically inactive. A comparison of ODCs with different structures shows that not only the size but also the position of the three-phase boundary is crucial for high electrode performance.
The dissertation is completed with a conclusion in the **seventh chapter**, which assesses the impact of the findings and provides an outlook for future work.

3 Measurements performed by Institute of Chemical and Electrochemical Process Engineering, Technische Universität Clausthal, Germany, published in [15].

2 Advanced Chlor-Alkali Electrolysis and Oxygen Depolarized Cathodes[4]

In this chapter, the processes within ODCs as well as the entire chlorine-alkali electrolysis are described based on the present literature. The aim of this chapter is to gather information about the system, the operating conditions and the processes taking place. This is essential to develop the model and to understand the interaction of processes, and the influence of process conditions. In the following, an overview of different chlor-alkali electrolysis processes – including the advanced electrolysis with ODCs – is given. Subsequently, the processes in ODCs are considered in detail. The essential findings of this review for the model design are summarized at the end of this chapter.

2.1 Processes of chlor-alkali electrolysis

In all conventional chlor-alkali electrolysis processes water and sodium chloride (NaCl) are used as reactants and chlorine (Cl_2) and caustic soda (NaOH) are obtained as products. Additionally, hydrogen is produced as a byproduct:

$$2\,\mathrm{NaCl} + 2\,\mathrm{H_2O} \rightarrow 2\,\mathrm{NaOH} + \mathrm{Cl_2}\uparrow + \mathrm{H_2}\uparrow \qquad E_{\mathrm{cell}} = 2.23\,\mathrm{V} \tag{2.1}$$

Where E_{cell} is the open circuit voltage (OCV) of the full cell under operating conditions.[5] In the novel energy-saving membrane process with ODCs, this reaction is changed. While oxygen is consumed as an additional reactant, hydrogen is not produced:

$$2\,\mathrm{NaCl} + \mathrm{H_2O} + 0.5\,\mathrm{O_2} \rightarrow 2\,\mathrm{NaOH} + \mathrm{Cl_2}\uparrow \qquad E_{\mathrm{cell}} = 1.01\,\mathrm{V} \tag{2.2}$$

---

4 Parts of this chapter have been published in Röhe et al., ChemSusChem, 2019 [1] and Röhe et al., ChemElectroChem, 2019 [2].

5 Voltages and potentials given in this chapter are referred to the operating conditions of the membrane process with and without ODC. Depending on the source, the values differ slightly due to varying process conditions. The data are taken from [6].

Due to the reduced cell voltage of this innovative process, the need for electrical energy for chlor-alkali electrolysis is reduced by about 30 % – in comparison to the most energy-efficient conventional process [6]. As discussed later, the energy savings due to more favorable thermodynamics are partly counteracted by the kinetic losses of the ODC.
In the following sub-sections, the conventional methods are introduced shortly, in-depth descriptions are given by Schmittinger et al [16]. Afterwards, the membrane process using ODCs is explained more in detail.

2.1.1 Conventional methods

In all three conventional processes – namely, the diaphragm process, the amalgam process and the membrane process – sodium chloride in an aqueous solution is used as the anolyte, where the NaCl is dissociated into its ions:

$$NaCl \rightleftharpoons Na^+ + Cl^- \tag{2.3}$$

As the actual anode reaction, chloride ions are oxidized to gaseous chloride:

$$2\,Cl^- \rightarrow Cl_2\uparrow + 2\,e^- \qquad E_0 = 1.33\,\text{V vs. SHE} \tag{2.4}$$

E_0 is the open circuit potential (OCP) of the half cell reaction.
At the cathode, the kinetically preferred reaction is the hydrogen evolution reaction (HER), wherein water is reduced to hydroxide ions and hydrogen [4]:

$$2\,H_2O + 2\,e^- \rightarrow H_2\uparrow + 2\,OH^- \qquad E_0 = -0.90\,\text{V vs. SHE} \tag{2.5}$$

In case that produced hydroxide ions get to the anode side, the formation of hypochlorite (ClO^-, eq. (2.6)) and chlorate (ClO_3^-, eq. (2.7)) take place as undesired side reactions:

$$Cl_2 + 2\,OH^- \rightarrow ClO^- + Cl^- + H_2O \tag{2.6}$$

$$6\,ClO^- + 3\,H_2O \rightarrow 2\,ClO_3^- + 4\,Cl^- + 6\,H^+ + 1.5\,O_2 + 6\,e^- \tag{2.7}$$

These side reactions would result in contamination of the product stream and in a reduced electricity yield [4]. The differences between the common chlor-alkali electrolysis techniques and the respective cathode reactions are mainly characterized by the way to ensure the absence of hydroxide ions in the anode chamber and thus to prevent the side reactions (2.6) and (2.7) [6].

In the diaphragm process, both half cells are separated by a porous diaphragm, which traditionally consists of asbestos. In newer applications, also PTFE and other inorganic substances are added to the asbestos [5]. At the cathode side, HER (eq. (2.5)) takes place in a caustic soda catholyte. To avoid hydroxide ions being transported to the anode, a convective flow of the anolyte is driven through the diaphragm to the cathode side. An advantage of this process is a relative low electrical energy demand of 2300 to 2900 $\mathrm{kWh\,t_{Cl_2}^{-1}}$ [6]. The main disadvantage is the sodium chloride contamination of the caustic soda solution [4].

In the amalgam process – also called mercury process – the alkaline HER and thus the formation of hydroxide ions is suppressed by the addition of mercury to the electrolyte. No separation between the anode and the cathode chamber is necessary. However, this process requires two separated electrochemical cells, the actual electrolysis cell and an amalgam decomposer. In the electrolysis cell the known reaction (2.4) takes place at the anode side. At the cathode, the Na^+ ions react with mercury to amalgam, no OH^- is formed [5]:

$$Na^+ + e^- + Hg_x \rightarrow NaHg_x \tag{2.8}$$

In the downstream decomposer, the mercury is recovered by electrochemical decomposition of the amalgam:

$$2\,NaHg_x + 2\,H_2O \rightarrow 2\,NaOH + H_2\uparrow + 2x\,Hg \tag{2.9}$$

This process requires with 3100 to 3400 $\mathrm{kWh\,t_{Cl_2}^{-1}}$ an increased amount of electrical energy [6], but a high purity of the caustic soda is obtained [4].

The membrane process combines the advantage of the low electrical energy demand (2100 to 2600 $\mathrm{kWh\,t_{Cl_2}^{-1}}$ [6]) of the diaphragm process and the high caustic soda purity of the amalgam process. In addition, no environmentally harmful substances such as asbestos or mercury are required [4].

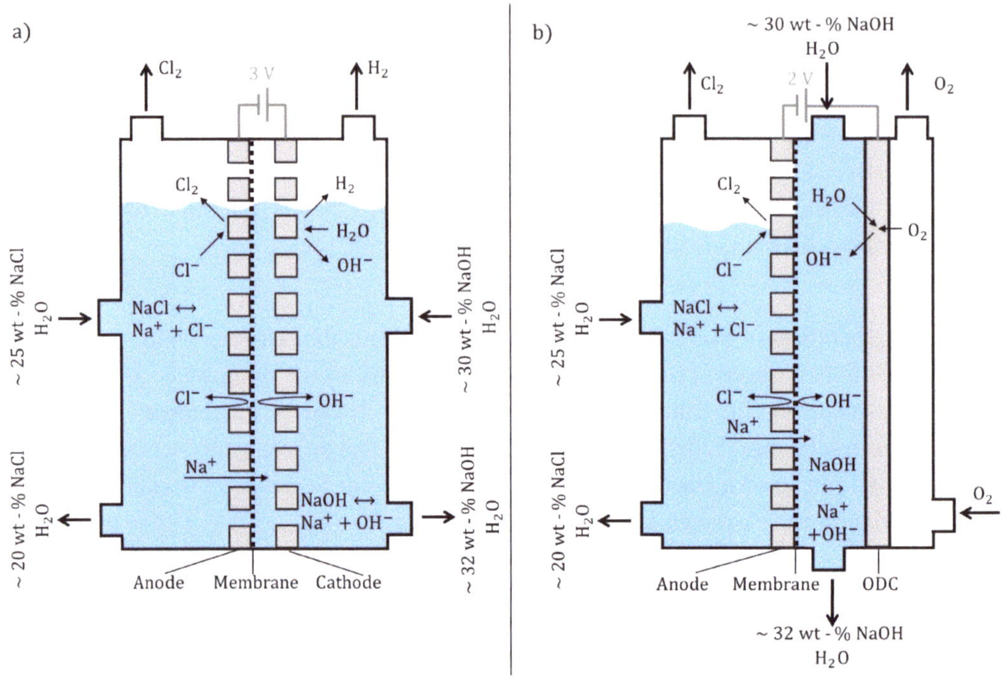

Figure 2.1: Schematic illustration of the chlor-alkali electrolysis membrane processes, a) without and b) with ODC, based on [3, 6].

As shown in fig. 2.1 a), the anode and the cathode side are separated by a membrane that is permeable to positive charged Na^+ but not to negative charged Cl^- and OH^-. Since the membrane must meet the requirements of long-term stability at high electrolyte concentrations and high temperatures, this process was a theoretical concept until about 1975; the realization was only made possible by the development of the Nafion™ membrane. To the anode chamber, a concentrated NaCl solution is continuously added. While the Cl^- reacts to chlorine gas (cf. eq. (2.4)), the Na^+ ions diffuse through the membrane into the cathode chamber. To the cathode side, a diluted sodium hydroxide solution is fed. At the cathode itself, the alkaline HER takes place (cf. eq. (2.5)). While the produced hydrogen is outgassing, the hydroxide ions are removed together with the sodium ions as concentrated caustic soda solution [6].

2.1.2 Membrane process with oxygen depolarized cathodes

The demand of electrical energy could be further reduced by revolutionizing the membrane process by utilization of ODCs [6]. Even though the advanced chlor-alkali electrolysis using ODCs was already theoretically presented by Butler in 1950 [17], the first plant in pilot scale was just set up in the 1990s [18]. Currently, it is already used industrially, inter alia in Germany and China [3]. As illustrated in fig. 2.1, compared to the conventional process, the cathode is replaced by an ODC, which is fed with the liquid electrolyte from the one side and with gaseous oxygen from the other side. The cathode reaction is changed from HER to ORR in alkaline milieu, where water reacts with oxygen to hydroxide ions:

$$O_2 + 2\,H_2O + 4\,e^- \rightarrow 4\,OH^- \qquad E_0 = 0.32\,\mathrm{V\ vs.\ SHE} \tag{2.10}$$

The advantage of this process lies in the cathodic cell potential: While in the conventional process the HER (eq. (2.5)) has a OCP of about $-0.90\,\mathrm{V}$ vs. Standard hydrogen electrode (SHE), the reversible cell potential is considerably higher for the oxygen reduction (eq. (2.10)) with $0.32\,\mathrm{V}$ vs. SHE. Although the overpotential of ORR is $\geq 0.7\,\mathrm{V}$ [18] and higher than for HER, under industrial operation conditions the total cell voltage can be reduced from about 3 V to about 2 V and therefore by about 30 %. Accordingly, the demand of electrical energy is reduced to about 1470 to $1820\,\mathrm{kWh\,t_{Cl_2}^{-1}}$. However, when considering the total energy consumption, it must be taken into account that oxygen with high purity is additionally required and the production of hydrogen is omitted [4].
A sustainability analysis of this process is presented by Jung et al. [19], deeper insights into the construction of industrial electrolyzers are given by Kintrup et al. [3]. An extensive overview of chlor-alkali electrolysis with ODC literature until 2008 is presented by Moussallem et al. [6].

2.2 Processes in oxygen depolarized cathodes

As illustrated in fig. 2.1 b), the ODC in an industrial chlor-alkali electrolyzer is fed with gaseous oxygen from the one side and with liquid electrolyte from the other side. At the gas-liquid interface within the ODC, oxygen dissolves into the liquid electrolyte and is further transported to the catalyst surface where the ORR (cf. eq. (2.10)) takes place. To enable the industrially required current densities of 4 to $6\,\mathrm{kA\,m^{-2}}$ [20], various demands must be ensured [18]:

- A catalytic active and electronically contacted electrode surface
- A high oxygen partial pressure in the gas phase
- A sufficient amount of oxygen dissolves into the electrolyte
- A satisfactory transport of water, oxygen and OH^- in the liquid phase

As a catalyst, typically silver is used. Silver is electronically highly conductive [18] and has a high catalytic activity for the ORR that is comparable to platinum when operating under industrial conditions [21] of NaOH-electrolyte concentrations exceeding 10 M and temperatures of 80 to 90 °C [6].
However, under these conditions the solubility of oxygen into the liquid electrolyte is poor [22]. To fulfill the criterion of a high dissolution rate of oxygen, the poor solubility must be compensated by a large size of the gas-liquid interface [18]. For a large interface between both phases, the proportions of pores filled with gas and pores filled with liquid electrolyte must be tared out well. To accomplish this, hydrophobic PTFE is used as second component in ODCs in addition to the hydrophilic silver. How different PTFE and silver loading influence the performance of an ODC is discussed by Franzen et al. [23] and Moussallem et al. [9].
The dissolution rate of oxygen into liquid electrolyte depends also on the partial pressure of oxygen in the gas phase at the gas-liquid interface. A high oxygen partial pressure is achieved by a high mass transport rate of oxygen in the gas phase. This implies that the evaporated water has to be removed through the gas-filled pores of the ODC [18].
In the liquid phase, the dissolved oxygen has to be transported from the gas-liquid interface to the catalyst surface. The mass transport resistance of this process is minimized, when the electrolyte forms a thin film of maximum a few micrometers on the catalyst surface. However, the nature of the liquid phase must also enable a sufficient transport of water as a reactant and hydroxide ions as the reaction product between the reactive zone and the liquid bulk phase [18].

At locations within the ODC where the requirements of a wetted, electrically connected catalyst, a nearby gas-liquid interface and sufficient mass transport in the gas and liquid phase are fulfilled, the ORR can take place. Although the reaction kinetics of ORR in the alkaline environment with a silver catalyst has been the subject of many studies [24–30], the exact reaction mechanisms are not clear [30], and various – partly contradictory – reaction paths are proposed. Several studies indicate that hydrogen peroxide may be formed as an intermediate reaction product [24–26]. But in general, silver is known to have a high catalytic activity towards peroxide reduction [31], so that a complete reduction of the oxygen and no hydrogen peroxide as a final product can be assumed. However, different reported reaction pathways coincide in the formation of the superoxide as the rate determining step [24–29], independent of pH value [26] or the crystal structure of the silver catalyst [25]:

$$O_2 + e^- \rightarrow O_2^- \tag{2.11}$$

Pinnow et al. gave the restriction that for ODCs under industrial conditions this assumption is not valid for cathodic overpotentials below 0.155 V [32].
Exemplary for a possible reaction path, the results of Ignaczak et al. [28] are discussed in the

following. Please note, even though this is the result of an extensive study, this mechanism is not validated for the specific process conditions of ODCs. In their recently released, comprehensive study combining density functional theory (DFT) with electrocatalysis theory and electron transfer theory, Ignaczak et al. postulats that the ORR in alkaline media on silver catalyst is proceeding in the steps below [28]:

$$O_2^- + H_2O + e^- \rightleftharpoons HO_2^- + OH^- \tag{2.12}$$

$$HO_2^- + H_2O \rightleftharpoons 2\,OH_{ad} + OH^- \tag{2.13}$$

$$2\,OH_{ad} + 2\,e^- \rightleftharpoons 2\,OH^- \tag{2.14}$$

According to the authors, the reaction steps (2.11) and (2.12) take place in the so-called outer-sphere mode [28], which means that the reactants do not adsorb on the catalyst, but are bound to the catalyst via a bridge of adsorbed OH^- and hydrogen bonds [33]. The O−O bond of the oxygen is split in the reaction step 2.13. For an in-depth discussion of the microkinetics, it is referred to the literature, e.g. Ignaezak et al. [28] or Ge et al. [30].
Further aspects of ODCs were inter alia studied by Botz et al., who measured an increasing ion concentration during ODC operation at the electrode surface [34]. Zhang et al. evaluated the electrochemistry of oxygen in concentrated NaOH-electrolyte [27].

2.3 Concluding remarks

In this chapter, literature about chlor-alkali electrolysis with ODCs and the ODC itself was reviewed. It was possible to gain insights into the electrode structure, the processes taking place as well as the process conditions.
Notwithstanding their great importance, ODCs as well as the interaction of processes and limitations in ODCs are not clearly understood: Properties like size, location and distribution of the gas-liquid interface are not identified yet, and the time constants of the different processes remain unknown. Also, it is not revealed yet, if the equilibrium of the gas-liquid interface stays constant during operation. This hinders knowledge-based improvement of ODCs. A promising tool to close these knowledge gaps is the development and use of dynamic models. As shown in various studies before, dynamic simulation gain information on the importance and interaction of mass transport, reaction mechanism and kinetics or geometrical structures. This is difficult to access in steady state simulation. Examples for complex model-based dynamic analysis are the identification of reaction kinetics [35] or the discrimination between charge transfer and mass transport

losses [36] using electrochemical impedance spectroscopy (EIS). A notable, tangible alternative is given by the evaluation of potential steps (multi-step chronoamperometry), chronoamperometry or potentiometry. These methods have been successfully used inter alia for identification of the transport and reaction processes in direct methanol fuel cells (DMFC) [37] or to investigate limiting factors for the ORR in metal-air batteries [12].
Based on the discussion in this chapter, the following points must be considered for the development of an ODC model:

- The model must be valid for typical industrial conditions such as high electrolyte concentrations exceeding 10 M, temperatures of 80 to 90 °C and current densities of 4 to 6 $\mathrm{kA\,m^{-2}}$.
- The model structure must reproduce a realistic gas-liquid interface.
- The phase transitions have to be considered, i.e. oxygen dissolution and water evaporation.
- In the gas phase, oxygen and water are the mainly transported species.
- In the liquid phase, the characteristic mass transport path of oxygen is from the gas-liquid interface to the catalyst surface. The characteristic mass transport path for water and ions is the distance between the reactive zone of the ODC and the liquid bulk.
- As various literature sources point out, the rate determining reaction step for the alkaline ORR is the formation of the superoxide (cf. eq. (2.11)). It must be verified whether this is also the case for the specific conditions of ODCs.

In order to develop a suitable ODC model, the above-mentioned aspects must be considered by choosing a feasible macroscopic modeling approach and mathematical-physical equations. That is the content of the following chapter.

3 Dynamic Modeling and Analyzing of Oxygen Depolarized Cathodes

In the previous chapter the characteristic processes in and properties of ODCs were discussed. It has been demonstrated that mass transport in the gas and the liquid phase, the dissolution of oxygen and evaporation of water are crucial factors for the performance of an ODC. For a knowledge-based improvement of ODCs, the influences as well as the interactions of these processes have to be understood, and the performance driver and limiting factors have to be identified. A suitable tool for this purpose is given by the dynamic modeling and analysis of ODCs.
In this chapter, it is demonstrated how the mentioned processes, their interactions as well as their contribution to the ODC performance can be analyzed with the help of dynamic macroscopic modeling. It is shown how the gas-liquid interface within the ODC can be represented in a macroscopic model and how ODCs have been modeled in previous studies using stationary methods. In the further course of the chapter, mathematical-physical approaches for modeling oxygen dissolution, water evaporation, reaction kinetics as well as mass transport in the liquid and gas phase are worked out specifically for the ODC. In addition, different dynamic electrochemical analysis techniques are presented which are suitable to evaluate the ODC and its processes by the aid of the macroscopic model. In the end, the chapter is concluded and an outline is given, which research questions will be answered in the further course of this dissertation using the dynamic model.

3.1 Mechanistic and literature models of oxygen depolarized cathodes

In this section, different mechanistic approaches to reproduce the gas-liquid interface of three-phase systems in porous structures are presented first. Second, an overview is given, how ODC have been modeled in previously publications.

3.1.1 Mechanistic models of gas diffusion electrodes

Different approaches of macroscopic modeling of porous GDE are reviewed by Yuh and Selman [38] or by Pinnow [39]. In the following the most important approaches are summarized and evaluated with regard to their ability to describe the gas-liquid interface sufficiently.

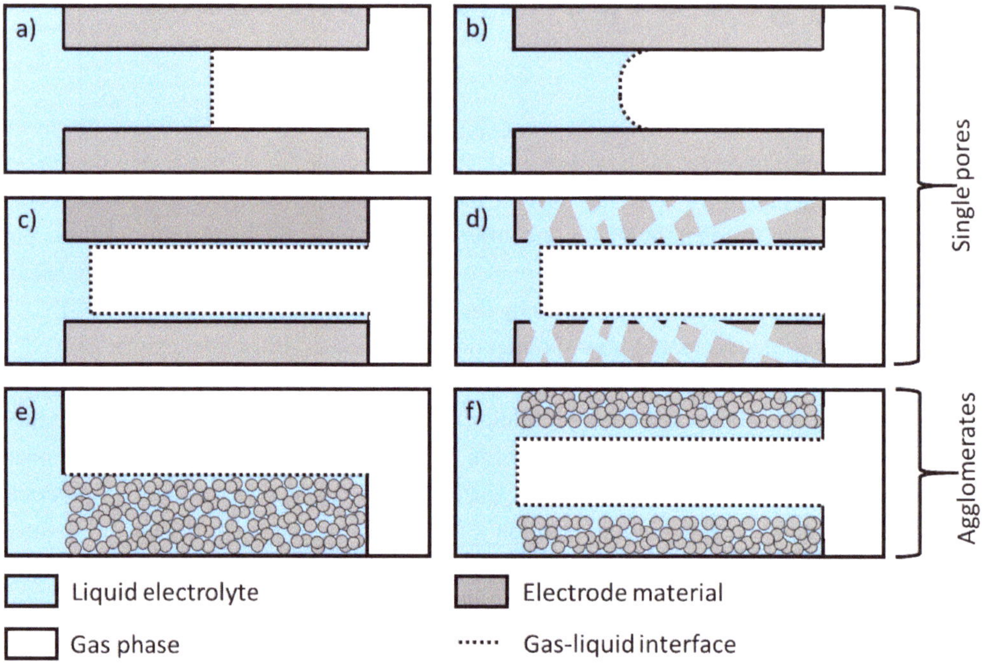

Figure 3.1: Schematic illustration of different model approaches for three-phase systems within a porous GDE: a) single-pore model (flooded), based on [38]; b) meniscus-model, based on [38]; c) thin-film model, based on [38]; d) model with micro and macro pores, based on [40]; e), flooded agglomerates model, based on [41] f); thin-film flooded agglomerate (TFFA) model, based on [42].

The first modeling approach is the simple-pore model, where a plane gas-liquid interface is assumed to be within or at the edge of a single pore, see fig. 3.1 a). That approach is characterized by a low gas-liquid interface which goes ahead with mass transport limitations at low current densities [43]. Extensions of this approach are the meniscus (fig. 3.1 b)) and the thin-film model (fig. 3.1 c)). In the meniscus model, it is assumed that the size of the gas-liquid interface within a pore is defined by the contact angle between the electrolyte and the pore wall. Dependent on the surface tension the meniscus is formed, which goes ahead with a higher gas-liquid interface [44]. In the thin-film model, the approach of an enlarge gas-liquid interface is taken further:

While the pore surface is wetted with a liquid thin-film, the inner pore space is filled with gas. Due to the enlarged gas-liquid interface, higher and more realistic limiting current densities can be modeled [43]. Typical values for the thickness of the thin-films are between 25 and 300 nm [45]. Burshtein et al. have presented a model which considers micro and macro pores (fig. 3.1 d)). While the larger macroscopic pores follow the thin-film approach, the smaller microscopic pores are filled with the liquid electrolyte. In contrast to the pure thin-film model, this approach is also able to satisfactorily model ionic mass transport and electrolyte conductivity. [40]
Separately, Giner and Hunter have developed an approach for a PTFE-bonded GDEs by changing the focus from single pores to larger structures (fig. 3.1 e)). According to their assumptions, the combination of a more hydrophilic catalyst and the hydrophobic PTFE creates macroscopic structures. These are represented by cylindrical flooded macroscopic pore networks of "flooded agglomerates" extending in the gas phase of the GDE. The flooded agglomerates themselves consist of microscopic pores filled with electrolyte. This approach has made it possible to simulate high ionic conductivities and current densities $j > 4\,\mathrm{kA\,m^{-2}}$ [41]. In order to reproduce mass transport more accurately, Cutlip has combined the thin-film with the flooded agglomerate approach to develop the thin-film and flooded agglomerate (TFFA) model (fig. 3.1 f)). The TFFA model considers flooded agglomerates within the porous GDE, which are additionally surrounded by a thin-film. The evaluation of the model has shown that the limiting current density is almost independent of the electrode thickness and the assumed shape of the flooded agglomerates [42].

3.1.2 Models of oxygen depolarized cathodes for chlor-alkali electrolysis

In literature, a high number of electrochemical models for oxygen reduction reaction in GDEs is known [46], but only a few address the specific process conditions of ODCs for chlor-alkali electrolysis. To the best of the author's knowledge, all these ODC models are stationary.
In 1997, Wang and Koda introduced their stationary model of an ODC as a TFFA approach with cylindrical flooded agglomerates [47]. In terms of mass transport, this model focuses exclusively on oxygen, while water and ions are not considered. The oxygen mass transport in the gas phase is calculated by Knudsen diffusion and in the thin-film as well as in the liquid phase by Fick's diffusion. The reaction kinetics of the ORR are modeled as a direct 4-e^--process and are implemented by an Tafel approach. The results demonstrate how a variation of different model parameters affects the polarization of ODCs. For example, a thicker thin-film leads to a lower limiting current density or a lower porosity leads to higher overpotentials independent of the current density [47]. The work of Wang and Koda was adapted by Sudoh et al. whereby the essential characteristics of the model remained unchanged [48]. By the means of the model,

experimental data of different ODCs were compared and analyzed. It was shown, how different specific catalyst surface areas or electrolyte intrusions affect the performance of ODCs [48]. The validity of the results of Sudoh et al. and Wang and Koda remain debatable, since important aspects such as water and ion mass transport were not considered.
The model of Wang and Koda was also revised and further developed by Pinnow et al. [32, 39]. Essential characteristics of the stationary pseudo two-dimensional TFFA model are Maxwell-Stefan diffusion in the gas and the liquid phase, a complete, homogeneous ORR as well as an oxygen solubility independent of the NaOH concentration of the liquid electrolyte. Especially the latter point is worth discussing, since Tromans determined a strong dependency of the oxygen solubility into the liquid electrolyte dependent on ion concentration [22]. In the model structure, a gas diffusion layer is considered at the interface to the porous electrode, but no liquid diffusion layer. Generally, it can be stated that the mass transport of water and hydroxide ions in the liquid electrolyte and their influence on the phase equilibrium seems to be underrepresented in the literature model by Pinnow et al. [32, 39]. The model predicts a fast oxygen mass transport in the gas phase and slow mass transport of the oxygen in the liquid phase which leads to a steep oxygen gradient within the thin-film. The limitation of the electric current is attributed to the depletion of oxygen in the liquid phase, which results mostly from the diffusion resistance of the oxygen cross the thin-film [32].
Although the dynamic one-dimensional three-phase model presented in this thesis was developed independently, the model structure, the considered processes as well as the determination of the material value and model parameters are based on the stationary pseudo-two-dimensional three-phase model by Pinnow et al. [32, 39].

3.2 Phase transitions

At the gas-liquid interface within ODCs, oxygen dissolution and water evaporation take place [8, 32]. As shown in different studies, the ion concentration in the liquid phase has a distinctive effect on oxygen solubility [22, 49, 50] and on water evaporation [5, 51, 52]. Botz et al. demonstrated that the NaOH concentration in ODCs can increase, depending on the current density, from 10 M at open circuit potential (OCP) conditions up to 20 M during operation [34]. Therefore, the consideration of ion concentration in both processes at the gas-liquid interface appears to be essential.
The solubility of oxygen in the electrolyte can generally be described by Henry's law:

$$c_{O_2} = \frac{p_{O_2}}{H_{O_2}} \tag{3.1}$$

Where c_{O_2} is the concentration of the dissolved oxygen in the liquid phase, p_{O_2} is the partial pressure of gaseous oxygen, and H_{O_2} is the inverse Henry constant. A calculation of H_{O_2} in NaOH electrolytes is proposed by Tromans [22]. Although the calculations are validated only up to 6 M, Zhang et al. were able to show that these calculations agree with experimental data up to 12 M [27]. Since no data validated beyond 12 M were found despite an extensive literature search, the calculations have to be further extrapolated in this thesis.

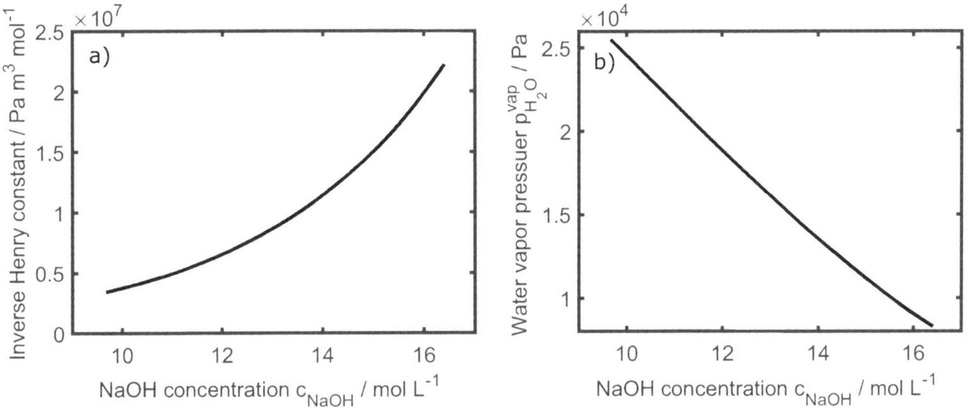

Figure 3.2: a) NaOH concentration dependent inverse Henry constant at 80 °C, calculated based on [22]; b) NaOH concentration dependent normalized water vapor pressure at 80 °C, calculated based on [51].

Fig. 3.2 a) shows the Henry constant for oxygen solubility at $T = 80\,°\mathrm{C}$ as a function of the NaOH concentration. In a 16 M electrolyte the oxygen solubility is 5 times lower than in a 10 M one. Using Henry's law with a oxygen partial pressure of 1 atm, the concentration of dissolved oxygen drops from 26.7×10^{-3} to $5.04 \times 10^{-3}\mathrm{mol\,m^{-3}}$ in the given interval. A comparison of further literature sources on oxygen solubility in the concentrated NaOH electrolytes is given by Pinnow [39].
The vapor pressure of water $p^{vap}_{H_2O}$ in concentrated NaOH solutions is documented by Hirschberg [51]. The normalized water vapor pressure over the ion concentration at $T = 80\,°\mathrm{C}$ is shown in fig. 3.2 b). While for a 10 M electrolyte the water partial pressure constitutes about 25 % of the gas phase pressure, the value drops to less than 10 % for a 16 M electrolyte.
In summary, both oxygen solubility and water vapor pressure decrease significantly with increasing NaOH concentration. Only small amounts of oxygen can be dissolved in the liquid electrolyte, but water partial pressure in the gas phase can reach 25 % of the total pressure. It is concluded that the partial pressure of water in the gas phase influences the availability of oxygen in the liquid phase, but the dissolved oxygen does not have a significant impact on the water

evaporation. However, the ion concentration dependencies of water evaporation and oxygen dissolution have to be considered in the model.

3.3 Reaction kinetics in oxygen depolarized cathodes

In literature studies, the reaction kinetic of ORR in a strong alkaline milieu has been modeled by its microscopic steps [24] or as lumped kinetic [29, 32].
A common way to calculate the reaction rate of ORR is the Butler-Volmer Equation (BVE) [46]. Detailed derivations are provided in the literature, e.g. by Bard and Faulker [53] or Hamann and Vielstich [4]. Herein, the BVE equation is adapted to the specific problem of the ODC, which is its applicability for multi-step reactions and simplifications due to large over-potentials.
The current density j_{F} for a general, reversible electrochemical reaction eq. (3.2), in which an oxidized species O reacts under the acceptance of n electrons to the reduced species R, can be described using Faraday's law, where the faradaic current density j_{F} is the sum of the oxidation j_{ox} and the reduction j_{ox} current densities [53].

$$O + n\mathrm{e}^- \underset{\mathrm{k_{red}}}{\overset{\mathrm{k_{ox}}}{\rightleftharpoons}} R \tag{3.2}$$

$$j_{\mathrm{F}} = j_{\mathrm{ox}} - j_{\mathrm{red}} = n\mathrm{F}(r_{\mathrm{ox}} - r_{\mathrm{red}}) \tag{3.3}$$

Where F is the Faraday constant and r_{ox} and r_{red} are the reaction rates. To describe the reaction rates of a multi-step reaction – as the ORR –, it has to be kept in mind, that the overall reaction rate of (electro-)chemical reactions is limited by its slowest, rate determining step (RDS). Thus, the kinetics of the RDS limits the total reaction rate: $r_{\mathrm{ox}} = r_{\mathrm{RDS,\,ox}}$ and $r_{\mathrm{red}} = r_{\mathrm{RDS,\,red}}$. The forward and backward rates of the RDS as single reaction step can be resolved according to the Butler-Volmer equation, depending on the overpotential η, the kinetic reaction rate constant $k_{0,\mathrm{red/ox}}$ and the activity $a_{\mathrm{RDS,red/ox},i}$ of the reactants involved in the RDS and ν as their stoichiometric coefficient [53]:

$$r_{\mathrm{RDS,ox}} = \prod_i a^{\nu}_{\mathrm{RDS,red},i} \cdot k_{0,\mathrm{ox}} \exp\left(\frac{\alpha \mathrm{F} z}{\mathrm{R}T}\eta\right) \tag{3.4}$$

$$r_{\mathrm{RDS,red}} = \prod_i a^{\nu}_{\mathrm{RDS,ox},i} \cdot k_{0,\mathrm{red}} \exp\left(-\frac{(1-\alpha)\mathrm{F} z}{\mathrm{R}T}\eta\right) \tag{3.5}$$

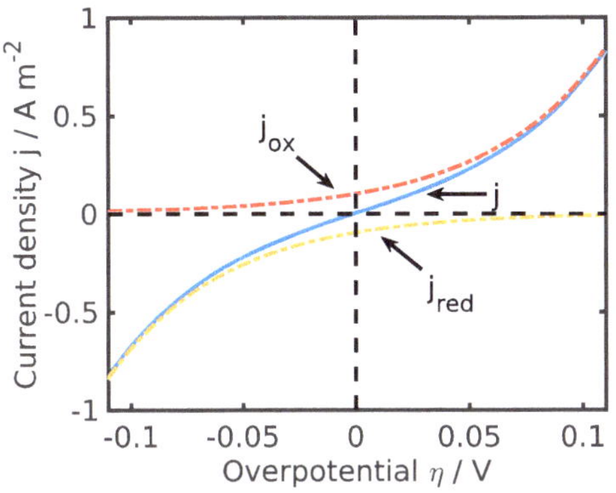

Figure 3.3: Dependency of current density on overpotential close to the OCP, calculated by Bulter-Volmer and Tafel equation. Following data are used: $\prod a^{\nu}_{\mathrm{RDS,red,i}} = \prod a^{\nu}_{\mathrm{RDS,ox,i}} = 1$; $n = z = 1$; $T = 298.15\,\mathrm{K}$; $k_{0,\mathrm{red}} = k_{0,\mathrm{ox}} = 0.1\,\mathrm{m^2\,mol\,s^{-1}}$; $\alpha = 0.5$.

Where α is the (anodic) symmetry factor, R the universal gas constant, T the temperature and z the transferred electrons in the RDS. Substituting the reaction rates into Faraday's law, the Butler-Volmer equation for a multi-step reaction is obtained as:

$$\begin{aligned} j_{\mathrm{F}} = n \cdot \mathrm{F} \prod_i a^{\nu}_{\mathrm{RDS,red},i} \cdot k_{0,\mathrm{ox}} \exp\left(\frac{\alpha \mathrm{F} z}{\mathrm{R}T}\eta\right) \\ - n \cdot \mathrm{F} \prod_i a^{\nu}_{\mathrm{RDS,ox},i} \cdot k_{0,\mathrm{red}} \exp\left(-\frac{(1-\alpha)\mathrm{F} z}{\mathrm{R}T}\eta\right) \end{aligned} \tag{3.6}$$

With this, the current density of an electrochemical multi-step reaction can be described with respect to the RDS. It should be noted that this only applies if the RDS is an electrochemical and not a chemical reaction [53]. It is also possible that the RDS may differ depending on the electrode potential [32] or on the activity of the reactants.

The current dependency due to the BVE on the overpotential is given in fig. 3.3, based on eqs. (3.3) - (3.6). The figure shows the total current density j_{F}, its oxidative j_{ox} and its reductive component j_{red} depending on the overpotential. As illustrated, at potentials near the OCP $-0.05\,\mathrm{V} \leq \eta \leq 0.05\,\mathrm{V}$, both partial currents have a considerable influence on the total current. At high positive overpotentials $\eta > 0.05\,\mathrm{V}$, the total current density j_{F} is governed by the oxidative current j_{ox}, while the reductive current j_{red} dominates at negative overpotentials $\eta < -0.05\,\mathrm{V}$. These properties can be used to simplify the BVE in case of high negative or positive overpotential. At high negative overpotentials $\eta \ll 0\,\mathrm{V}$, in eq. (3.5), $\exp\left(-\frac{(1-\alpha)\mathrm{F}z}{\mathrm{R}T}\eta\right)$ becomes a high value, while $\exp\left(\frac{\alpha \mathrm{F}z}{\mathrm{R}T}\eta\right)$ of eq. (3.4) approximates zero. Thus, at high negative overpotential,

j_{ox} can be neglected and the Butler-Volmer equation can be reduced to the so-called Tafel equation, where $j = -j_{\text{red}}$. For high positive overpotentials $\eta \ll 0\,\text{V}$ it works the other way round. However, this leads to the fact that the description of the reaction rate loses its validity near the OCP [53]. Finally, the current density of the ORR under the given conditions as an electrochemical multi-step reaction at higher negative overpotentials $\eta < -0.05\,\text{V}$ can be simplified to the Tafel equation, which takes the total number of electrons transferred and the reaction rate through the RDS into account:

$$j = -n \cdot \text{F}k_0 \cdot \prod_i a^{\nu}_{\text{RDS,ox},i} \exp\left(-\frac{(1-\alpha)\text{F}z}{\text{R}T}\eta \right) \tag{3.7}$$

The overpotential used in the BVE and the Tafel equation is defined as the difference between the potential in the reaction zone E and the open circuit potential of the electrode E_0:

$$\eta = E - E_0 \tag{3.8}$$

The open circuit potential results from the thermodynamics of the electrode and describes the cell potential at a net current flow of $j = 0\,\text{A}\,\text{m}^{-2}$. E_0 can be calculated from the tabulated value of the standard electrode potential E_{00} corrected by the activities $a_{i,\text{ox,red}}$ of the oxidized or reduced species by the Nernst equation [4]:

$$E_0 = E_{00} - \frac{\text{R}T}{n\text{F}} \ln\left(\frac{\prod_i a_{i,\text{ox}}^{|\nu_i|}}{\prod_i a_{i,\text{red}}^{|\nu_i|}} \right) \tag{3.9}$$

n is the number of electrons transferred in the overall reaction. Regardless of the RDS, in a multi-step reaction the activities of the total reaction are used in the Nernst equation to determine the OCP. In contrast to the kinetics, not only the activities of the reactants but also the activities of the products influences the thermodynamics of the reaction [53].

3.4 Mass transport in oxygen depolarized cathodes

In the gas phase of ODCs, water and oxygen are transported. The liquid phase is characterized by multi-component mass transport of water, hydroxide ions and the dissolved oxygen. The complexity is further increased by the phase transition, the electrochemical reactions and the high ion concentration of the liquid electrolyte.

In an electrochemical system, a distinction between three different mechanisms of mass transport is made [53]:

1. **Diffusion:** Mass transport due to a gradient in chemical potential, in idealized systems also due to a gradient in concentration or partial pressure.

2. **Convection:** Mass transport due to a gradient in density (natural convection) or as a forced convection due to pressure gradients. Additional convection can apply due to reactions or phase transitions in form of the so called Stefan-flow [54].

3. **Migration:** Mass transport of charged species due to a gradient in electrical potential.

In a binary NaOH-electrolyte, due to the constraint of electroneutrality, the electrical potential or rather the gradient of electrical potential can be eliminated as driving force from the mass transport equation. Therefore the migration can be neglected [55] and is not considered further.
For the diffusion process it must be distinguished between two basically different approaches: Fick's diffusion and Maxwell-Stefan diffusion. In contrast to Maxwell-Stefan diffusion, Fick's diffusion is based on the assumption of an idealized system [56]. Moreover, Fick's approach can only be transferred to multi-component diffusion in concentrated solutions to a limited extent [57, 58].
The bi-mixture of oxygen and water in the gas phase can be idealized in terms of mass transfer, thus the gradient in partial pressure can be used to calculate the diffusion flux. An additional convection may occur due to the oxygen dissolution and water evaporation. This superposing convection is considered as the Stefan-flow. The mass transport in the liquid electrolyte is characterized by high ion concentrations and thus a strictly non-ideal behavior, multi-component mass transport as well as phase transitions. Consequently, these properties are considered by the Maxwell-Stefan diffusion in combination with the Stefan-flow in the herein developed model.
In the following, the Maxwell-Stefan equation for the specific system of ODCs is derived. Subsequently, the Stefan-flow is described on the example of Fick's diffusion.

3.4.1 Maxwell-Stefan diffusion

In this chapter, the dynamic Maxwell-Stefan relation for the specific system of the liquid phase within an ODC is derived beginning from the ordinary Maxwell-Stefan diffusion for the stationary case. Herein, the diffusion is discussed exclusively; the species balance, including sinks and sources, are considered in chapters 4, 5 and 6.
For a better understanding of the Maxwell-Stefan relation, the derivation of the Maxwell-Stefan diffusion for the stationary case is attached in appendix A and briefly summarized hereafter.

As illustrated in fig. 3.4, the basic idea of the Maxwell-Stefan diffusion is that the drag force of species i based on the gradient in its chemical potential is equal to the forces acting on species i due to its relative velocity to the surrounding species j $(i \neq j)$ [59, 60].

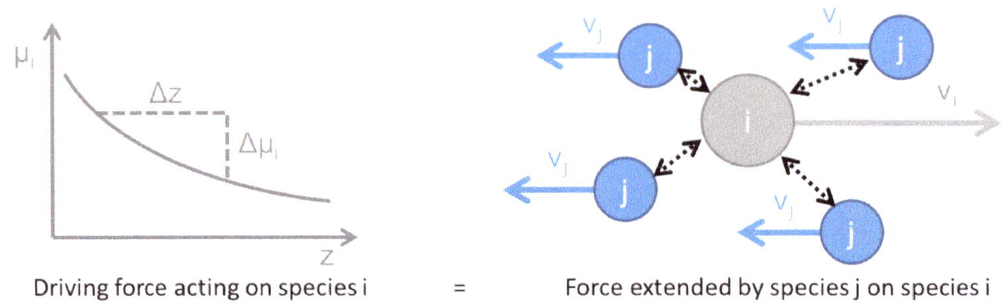

Figure 3.4: Schematic illustration of Maxwell-Stefan diffusion for the binary case. Partly based on [56].

The drag force $d_i \cdot \mathrm{R}T \cdot c_{\mathrm{total}}$ acting on species i due to its gradient in chemical potential is given by eq. (3.10) [60] (cf. A.11).

$$d_i \cdot \mathrm{R}T \cdot c_{\mathrm{total}} = x_i \frac{\mathrm{d}\mu_i}{\mathrm{d}z} \cdot c_{\mathrm{total}} \tag{3.10}$$

Where c_{total} is the total concentration of all species, $\frac{\mathrm{d}\mu_i}{\mathrm{d}z}$ is the gradient in the chemical potential, x_i is the mole fraction and d_i is a coefficient for the forces acting on the molecules of species i.
The force $d_i \cdot \mathrm{R}T \cdot c_{\mathrm{total}}$ which is applied on the molecules of one species i due to their relative velocity to the other present species j is calculated by eq. (3.11) [59] (cf. A.10).

$$d_i \cdot \mathrm{R}T \cdot c_{\mathrm{total}} = -\sum_{\substack{j=1\\ i\neq j}}^{n} \frac{x_i x_j \cdot (v_i - v_j)}{Ð_{i,j}} \cdot \mathrm{R}T \cdot c_{\mathrm{total}} \tag{3.11}$$

Here $Ð_{i,j}$ is the binary Maxwell-Stefan diffusion coefficient of species i and j, and v_i the average velocity of species i.
The Maxwell-Stefan diffusion for the stationary case is obtained by equaling both forces given in eq. 3.10 and 3.11 and using the relation $c_i = x_i \cdot c_{\mathrm{total}}$ (cf. A.12):

$$\frac{c_i}{\mathrm{R}T} \frac{\mathrm{d}\mu_i}{\mathrm{d}z} = -\sum_{\substack{j=1\\ i\neq j}}^{n} \frac{c_j c_i \cdot (v_j - v_i)}{c_{\mathrm{total}} Ð_{i,j}} \tag{3.12}$$

The transformation of the stationary Maxwell-Stefan diffusion into the dynamic form begins with the calculation of the molar flow densities: The net flux $\dot{N}_i$ of the species i can be calculated by [59]:

$$\dot{N}_i = -\sum_{\substack{j=1 \\ i \neq j}}^{n} Ð_{i,j} \cdot c_{\text{total}} \cdot d_{i,j} \tag{3.13}$$

By substituting $d_{i,j}$ by the right hand side of equation 3.11, we get:

$$\dot{N}_i = -\sum_{\substack{j=1 \\ i \neq j}}^{n} Ð_{i,j} \cdot c_{\text{total}} \frac{x_i x_j \cdot (v_i - v_j)}{Ð_{i,j}} \tag{3.14}$$

which can be simplified to

$$\dot{N}_i = -\sum_{\substack{j=1 \\ i \neq j}}^{n} \frac{c_i c_j \cdot (v_i - v_j)}{c_{\text{total}}} \tag{3.15}$$

Now the resulting flux $\dot{N}_i$ of the species i is given as a function of the relative velocities $(v_i - v_j)$ causing momentum exchange between different species and the frequency of the collisions, represented by $c_i c_j$. That equation gets clarified by noting that the Maxwell-Stefan relation only allows $n-1$ independent fluxes for n interacting species [59]. The velocities v_i and v_j depend on the driving forces of the species. Since in the binary NaOH-electrolyte the migration can be neglected and convection only may occur as a Stefan-flow, which leads to a shift of the system and not to a change of the relative velocity $(v_i - v_j)$ between the different species, an exclusive diffusion as driving forces for v_i and v_j can be assumed. For an exclusive diffusion in a concentrated electrolyte, the velocity v_i can be calculated by eq. 3.16 [60] The calculation of v_j is done analogously.

$$v_i = \frac{Ð_{i,j}}{\mathrm{R}T} \cdot \frac{\mathrm{d}\mu_i}{\mathrm{d}z} \tag{3.16}$$

Please note, that eq. (3.16) and all following steps would change, if e.g. migration-based fluxes respective to the surrounding medium would occur. An overlaying convective flow, which moves as species with the same velocity and direction is still valid. From eq. (3.15) and (3.16), eq. (3.17) is obtained.

$$\dot{N}_i = -\sum_{\substack{j=1 \\ i \neq j}}^{n} \frac{Ð_{i,j}}{\mathrm{R}T} \frac{c_i c_j}{c_{\text{total}}} \left(\frac{\mathrm{d}\mu_i}{\mathrm{d}z} - \frac{\mathrm{d}\mu_j}{\mathrm{d}z} \right) \tag{3.17}$$

For the diffusive mass transport, the change of concentration depending on the time equals the derivation of the flux $\dot{N}_i$ over the location z:

$$\frac{\mathrm{d}c_i}{\mathrm{d}t} = \frac{\mathrm{d}\dot{N}_i}{\mathrm{d}z} \tag{3.18}$$

Using eq. (3.17) and (3.18), the change of the concentration depending on time can be calculated. Since c_i, c_j, μ_i, μ_j and c_{total} are depending on the location, the total differential of (3.18) is needed, which is given in eq. (3.19). A constant diffusion coefficient $\frac{\partial Đ_{i,j}}{\partial z} = \frac{\partial Đ_{i,j}}{\partial t} = 0$ is assumed.

$$\begin{aligned}\frac{\mathrm{d}c_i}{\mathrm{d}t} = \sum_{\substack{j=1\\ i\neq j}}^{n} \frac{Đ_{i,j}}{\mathrm{R}T}\Bigg(&\frac{\partial^2\mu_i}{\partial z^2}\cdot\frac{c_ic_j}{c_{\text{total}}} - \frac{\partial^2\mu_j}{\partial z^2}\cdot\frac{c_ic_j}{c_{\text{total}}}\\ &+\frac{\partial\mu_i}{\partial z}\cdot\frac{\partial c_i}{\partial z}\frac{c_j}{c_{\text{total}}} - \frac{\partial\mu_j}{\partial z}\cdot\frac{\partial c_i}{\partial z}\frac{c_j}{c_{\text{total}}}\\ &+\frac{\partial\mu_i}{\partial z}\cdot\frac{\partial c_j}{\partial z}\frac{c_i}{c_{\text{total}}} - \frac{\partial\mu_j}{\partial z}\cdot\frac{\partial c_j}{\partial z}\frac{c_i}{c_{\text{total}}}\\ &+\frac{\partial\mu_i}{\partial z}\cdot\frac{c_ic_j}{c_{\text{total}}^2}\frac{\partial c_{\text{total}}}{\partial z} - \frac{\partial\mu_j}{\partial z}\cdot\frac{c_ic_j}{c_{\text{total}}^2}\frac{\partial c_{\text{total}}}{\partial z}\Bigg)\end{aligned} \tag{3.19}$$

As discussed in chapter 4 and appendix B in the range of industrial relevant working conditions for the advanced chlor-alkali electrolysis, the total concentration c_{total} of the liquid phase can be handled as constant. With a constant c_{total}, eq. (3.19) can be simplified to:

$$\begin{aligned}\frac{\mathrm{d}c_i}{\mathrm{d}t} = \sum_{\substack{j=1\\ i\neq j}}^{n} \frac{Đ_{i,j}}{\mathrm{R}T}\Bigg(&\left(\frac{\partial^2\mu_i}{\partial x^2} - \frac{\partial^2\mu_j}{\partial z^2}\right)\cdot\frac{c_ic_j}{c_{\text{total}}}\\ &+\left(\frac{\partial\mu_i}{\partial z} - \frac{\partial\mu_j}{\partial z}\right)\cdot\frac{\partial c_i}{\partial z}\frac{c_j}{c_{\text{total}}}\\ &+\left(\frac{\partial\mu_i}{\partial z} - \frac{\partial\mu_j}{\partial z}\right)\cdot\frac{\partial c_j}{\partial z}\frac{c_i}{c_{\text{total}}}\Bigg)\end{aligned} \tag{3.20}$$

Equation (3.20) describes the time-dependent change in concentration of species i, respecting interaction with all species $j \neq i$ and assuming a constant total concentration. The driving force is exclusively diffusion, based on gradients in chemical potential.

A simplification of the equation (3.20) can be achieved, if an ideal behavior of the involved species is assumed. Using activity coefficients $\gamma_i = 1$, the concentration is obtained as the driving force instead of the chemical potential. As explained in detail in appendix A, in this case equation (3.20) can be simplified to equation (3.21):

$$\frac{\mathrm{d}c_i}{\mathrm{d}t} = \sum_{\substack{j=1 \\ i \neq j}}^{n} Đ_{i,j} \left(\frac{\partial^2 c_i}{\partial z^2} \frac{c_j}{c_{\text{total}}} - \frac{\partial^2 c_j}{\partial z^2} \frac{c_i}{c_{\text{total}}} \right) \tag{3.21}$$

Eq. (3.21) is the Maxwell-Stefan diffusion which regards the interaction between the species i and j and the gradient in concentration as the driving force. All further before mentioned assumptions are still valid.
With the aid of the equations (3.20) or (3.21) developed here, multi-component mass transport in the liquid phase can be described dynamically for the specific system of ODCs.

3.4.2 Fick's diffusion and Stefan-flow

In the gas phase of an ODC, oxygen and water are present, which can be assumed to behave as ideal gases. Thus the diffusion can be calculated by Fick's law. The aim of this section is to expand the pure Fick's diffusion by the overlaying convective Stefan-flow, which may occur in ODCs as a result of the water evaporation and oxygen dissolution. Herein, Fick's first law is taken as given.
Dependent on the properties of a system, three different mechanisms of diffusion can occur [54]:

1. The bilateral equimolar diffusion
2. The bilateral non-equimolar diffusion
3. The unilateral diffusion

In the first case, the bilateral equimolar diffusion, the number of molecules moving in opposite directions are equal to each other. This mechanism can be illustrated by two closed vessels connected by a tube. The vessel 1 is mostly filled with the gas A, the vessel 2 is mostly filled with the gas B, no gradients in temperature or pressure are present (see fig. 3.5 a)) [54].

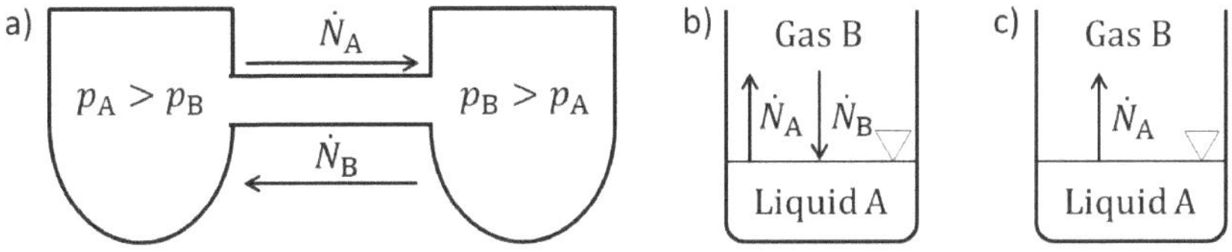

Figure 3.5: Schematic illustration of a) bilateral equimolar diffusion (based on [54]); b) bilateral non-equimolar diffusion; c) unilateral diffusion.

In an idealized system, it can be assumed that gas A diffuses through the connecting tube towards vessel 1 and gas B diffuses through the connecting tube towards vessel 2. The fluxes of both gases can be calculated by the mean of Fick's first law, with $\dot{N}_A$ and $\dot{N}_B$ as the fluxes of the respective species, p_A and p_B as the partial pressures, z as the location and $\mathscr{D}_{A,B}$ as the Fick's diffusion coefficient: [54]

$$\dot{N}_A = -\frac{\mathscr{D}_{A,B}}{RT}\frac{dp_A}{dz} \tag{3.22}$$

$$\dot{N}_B = -\frac{\mathscr{D}_{A,B}}{RT}\frac{dp_B}{dz} \tag{3.23}$$

Since no gradient in total pressure occurs, $p_A + p_B = p_{total} = const.$, this results in $\frac{dp_A}{dz} = -\frac{dp_B}{dz}$. Further taking $\mathscr{D}_{A,B} = \mathscr{D}_{B,A}$ into account, it becomes obvious that $\dot{N}_A = -\dot{N}_B$ and that the number of the transported molecules of both species are the same [54]. This is not the case for bilateral non-equimolar diffusion.

The bilateral non-equimolar diffusion is particularly important when the diffusion process is associated with sinks and sources. Examples for sinks and sources can be (electro-)chemical reactions or phase transitions. The mechanism is illustrated by the example of the phase transition. Here, a model system can be a beaker half filled with a liquid A and half with a gas B, whereby the liquid evaporates into the gas phase and the gas dissolves into the liquid (see fig. 3.5 b)). It can be expected that the flux of the evaporating liquid $\dot{N}_B$ and the flux of dissolving gas $\dot{N}_A$ will be unequal in amount. The bilateral non-equimolar diffusion is based on the assumption, that the inequality of both diffusion fluxes is compensated by superimposed convective fluxes $v_s \cdot p_A$ and $v_s \cdot p_B$, the so-called Stefan-flow, with v_s as the velocity of the convective mass transport. The fluxes of both species can not longer be described by a pure diffusive term as in eq. (3.22) and eq. (3.23), but also with a convective term: [54]

$$\dot{N}_A = -\frac{\mathscr{D}_{A,B}}{RT}\frac{dp_A}{dz} + v_s \cdot \frac{p_A}{RT} \tag{3.24}$$

$$\dot{N}_B = -\frac{\mathscr{D}_{A,B}}{RT}\frac{dp_B}{dz} + v_s \cdot \frac{p_B}{RT} \tag{3.25}$$

Here again, at a constant pressure and temperature, the absolute values of the actual diffusion fluxes of both components are equal to each other: $\mathscr{D}_{A,B}\frac{dp_A}{dz} = -\mathscr{D}_{B,A}\frac{dp_B}{dz}$. In contrast, the convective Stefan-flow and thus the total fluxes are unequal, so that $\dot{N}_A \neq -\dot{N}_B$. The velocity of the Stefan-flow v_s has to be determined – according to the system – from the fluxes across the interface or the stochiometry of the reaction. The change in partial pressure of the species dependent on the time can be determined by eq. (3.18) and is given for the general case with $v_s = const.$, as [54]:

$$\frac{\partial p_i}{\partial t} = \mathscr{D}_{i,j}\frac{\partial^2 p_i}{\partial z^2} + v_s\frac{\partial pi}{\partial z}. \quad (3.26)$$

Since the Stefan-flow is a shift of the whole system (same velocity and flow direction for all components), the approach of the Stefan-flow can also be combined with the Maxwell-Stefan diffusion.

The last case, the unilateral diffusion, represents a special case of the bilateral non-equimolar diffusion. Examples are given by a two-gas mixture in a chamber sealed with a selective membrane, or by a two phase system with either evaporation or dissolution (see fig. 3.5 c)) [54]. However, this case is not considered further here, since this mechanism is not expected in ODCs.

3.5 Dynamic analysis of oxygen depolarized cathodes

The performance of ODCs is determined by a complex interplay of its reaction kinetics, mass transport in gas phase and mass transport in liquid phase which is difficult to access in steady state simulation. One way to separate the influences of the different processes on electrode performance from each other is the utilization of dynamic methods [35]. The principle of dynamic methods is based on the excitation of the system by a time-dependent change of at least one process parameter (e.g. potential, current, temperature). From the time-dependent system response, inter alia, the characteristic time constants τ of the excited processes can be determined. In fig. 3.6 different processes which may take place within ODCs and also suitable dynamic electrochemical methods to study them are assigned to their characteristic time constants.

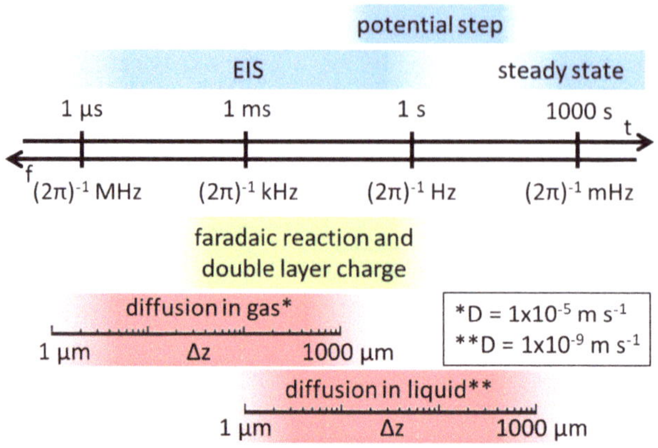

Figure 3.6: Time constants of processes in ODCs and suitable methods to detect them, Δz is the characteristic diffusion length, partly based on [13, 35].

The time constants of the processes within ODCs, can range from milliseconds to several minutes. In ODCs diffusion occurs in gas and in liquid phase. For both cases, the time constant is significantly increasing with the diffusion length Δz. But for the same diffusion length, the time constant of gas and liquid diffusion are magnitudes apart. This can be explained by the equation of the diffusion time constant, which is derived from Fick's second law [13]:

$$\tau_{\mathrm{mt}} \approx \frac{\Delta z^2}{D} \tag{3.27}$$

Where τ_{mt} is the time constant of the diffusion process, D is the diffusion coefficient and Δz is the mass transport length. Here it can be seen that the time constant increases quadratically with the diffusion length and decreases proportional with the diffusion coefficient. Typical values for the gas phase diffusion coefficients are about 1×10^{-6} to $1 \times 10^{-4}\,\mathrm{m^2\,s^{-1}}$ and for the liquid diffusion about 1×10^{-10} to $1 \times 10^{-8}\,\mathrm{m^2\,s^{-1}}$, which explains the strongly different time constants of diffusion in the two phases [61].

Another characteristic time constant is received due to the charge and discharge of the double layer. This time constant can differ depending on further electrode processes. For example has the time constant of double layer charge parallel to faradaic reactions to be calculated differently than the time constant of double layer charge of a blocked electrode surface without faradaic reaction. For a parallel reaction – as expected in ODCs – the following equation is given [62]:

$$\tau_{\mathrm{dl}} = C_{\mathrm{dl}} \cdot R_{\mathrm{ct}} \tag{3.28}$$

With the characteristic time constant τ_{dl}, the double layer capacitance C_{dl} and the charge transfer resistance R_{ct}. The charge transfer resistance is a complex parameter that is determined by the reaction kinetics of the electrode. The double layer capacitance can be effected by many factors, such as catalyst loading, ion concentration in the electrolyte or applied potential [4].
Additional processes which may be dynamic with characteristic time constants are e.g. the heat conduction for non-isothermal systems [13], or degradation processes. Öhl et al. showed that degradation of the silver catalyst in ODCs progress in the range of weeks [63] and is therefore negligible for this work.

Fig. 3.6 also gives an overview of suitable methods to analyze the processes depending on their time constants. Herein, dynamic methods are focused on the electrochemical impedance spectroscopy (EIS) and the potential steps. An overview about further dynamic electrochemical methods and their field of application are provided by Bard and Faulker [53] or Hamann and Vielstich [4]. While for faster time constants EIS is a suitable, high resolution tool, potential steps are feasible to detect slower time constants, especially for diffusion processes and cover a wider range of time constants. Electrochemical impedance spectroscopy has been used successfully e.g. for identification of reaction kinetics in DMFC [35], or to analyze catalyst poisoning in proton exchange membrane fuel cell (PEMFC) cathodes [64]. By applying potential steps on DMFC, the influences of kinetics and transport have been examined [37].
The potential step – also called (multi-)step chronoamperometry – is an instationary method, in which the electrode potential is changed abruptly at a defined time as input signal. The time-dependent current density is obtained as output signal [53]. Input, output and the triggered processes within the ODC are illustrated in fig. 3.7.

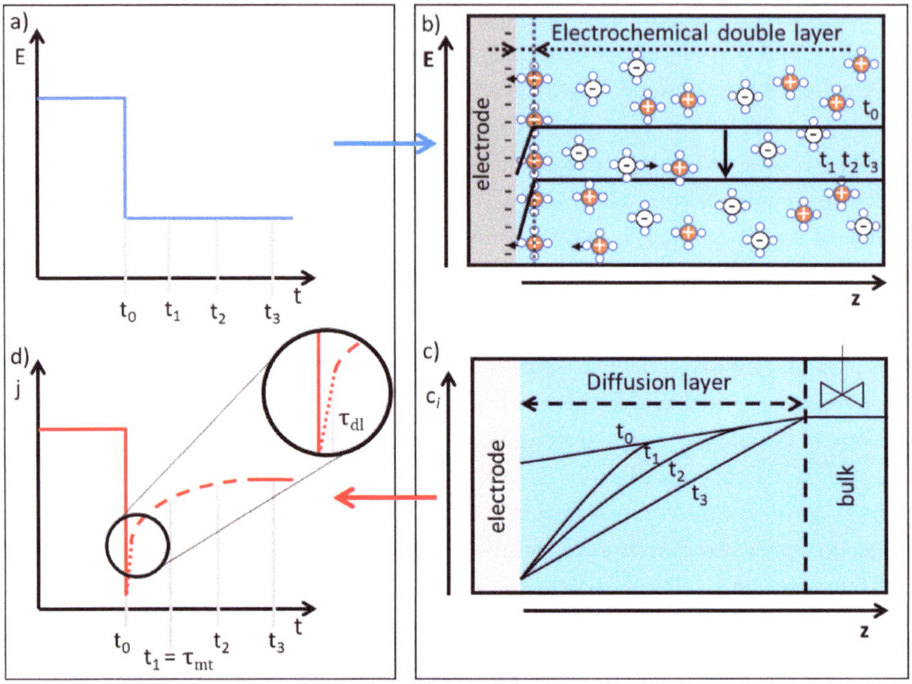

Figure 3.7: Schematic illustration of changes in states during a potential step: a) potential dependent on time as input signal; b) double layer charge / discharge (illustration for ion concentrations $\geq$ 1 M) as fast process; c) mass transport due to electrochemical surface reactions as slow process and d) current density dependent on time as output signal. Partly based on [4, 53].

The change in potential (fig. 3.7 a)) triggers the double layer charge (fig. 3.7 b)), the faradaic reactions and thus the mass transport (fig. 3.7 c)). Other processes may also be stimulated, but will not be considered further here. The current density (fig. 3.7 d)) is characterized by a peak immediately after the step, followed by a slower relaxation. The peak is assigned to the fast process of the double layer charge. The subsequent relaxation is the result of a slower diffusion process. The time constants of the double layer charge / discharge τ_{dl} and the diffusion τ_{mt} can be inferred directly from the time-dependent response of the current.

As also shown in fig 3.6, each characteristic time constant is associated with a characteristic frequency f_{char}. The time-behavior of the individual processes can be transferred into the frequency domains by:

$$f_{\text{char}} = \frac{1}{2\pi\tau} \tag{3.29}$$

The quasi-stationary EIS makes use of this relationship. Herein, a short overview about EIS is given, deeper insight into the theory of EIS is presented by Orazem and Tribolet [62], the effects of different processes on the impedance spectra are evaluated by McDonald [65]. The general principle of EIS is sketched in fig. 3.8. The electrochemical system is stimulated by a sinusoidal potential E with the frequency f and a small amplitude, the sinusoidal current density j is considered as the output signal [4]. At a certain frequency, the processes with the associated characteristic frequency are triggered, cf. fig. 3.8 a) - d). To excite different processes according to their specific frequency the procedure is repeated for several orders of magnitudes of frequencies.

For each frequency f, the impedance magnitude $|Z(f)|$ and the phase shift $\varphi(f)$ is determined. As illustrated in 3.8 e), the phase shift $\varphi(f)$ is obtained from the time shift $\Delta t(f)$ between the sinusoidal curve of the potential and the sinusoidal curve of the current.

$$\varphi(f) = \Delta t(f) \cdot 2\pi f \tag{3.30}$$

The impedance magnitude $|Z(f)|$ is calculated from the amplitude of the sinusoidal potential $\hat{E}$ and the sinusoidal amplitude of the current $\hat{J}(f)$ [62].

$$|Z(f)| = \frac{\hat{E}}{\hat{J}(f)} \tag{3.31}$$

Based on $|Z(f)|$ and $\varphi(f)$, the impedance can be divided into its real $Z_\mathrm{R}(f)$ and imaginary parts $Z_\mathrm{I}(f)$:

$$Z_\mathrm{R}(f) = \cos(\varphi(f)) \cdot |Z(f)| \tag{3.32}$$

$$Z_\mathrm{I}(f) = \sin(\varphi(f)) \cdot |Z(f)| \tag{3.33}$$

One way of presenting the resulting spectra is the Nyquist plot, in which usually the negative imaginary part of the impedance for all frequencies $-Z_\mathrm{I}$ is plotted over the real part Z_R. In fig. 3.8 f) a Nyquist plot for a electrochemical system with both processes, double layer charge / discharge parallel to a faradaic reaction as well as mass transport, is illustrated. Please note that the contribution of a process to the electrode dynamics can be negligibly small, in such a case the process will not be visible in the Nyquist plot. In the Nyquist plot, each process is usually represented by a semi-circle. It should be mentioned that other shapes, such as a Warburg line with a 45° slope, are possible; however, since these are not expected in an ODC, here they will not be considered any further. Two essential characteristics of the semi-circles are their diameters and their characteristic frequencies. The characteristic frequency f_char is the frequency at which the imaginary value of the semi-circle reaches its maximum. The diameter of

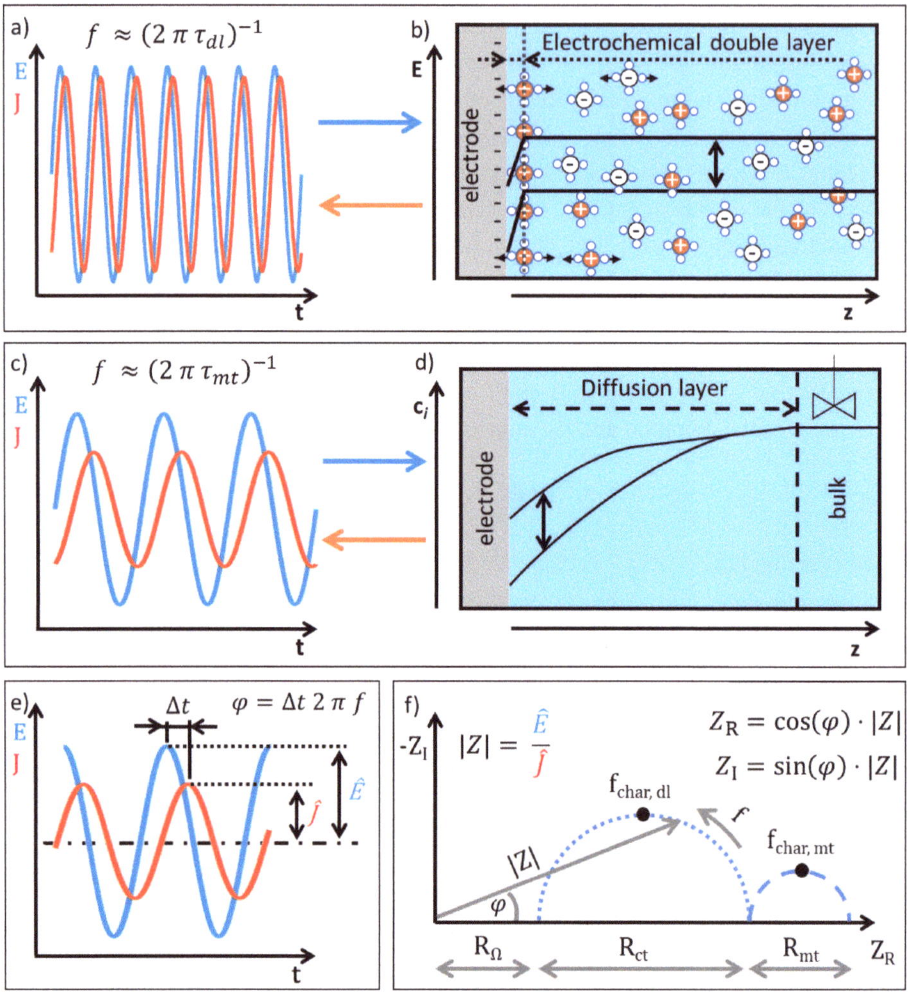

Figure 3.8: Schematic illustration of electrochemical impedance spectroscopy analysis and the related dynamic state changes: a) sinusoidal potential as input signal and current as output signal in high frequency range; b) double layer charge / discharge (illustration for ion concentrations $\geq 1\,\mathrm{M}$) as fast process; c) sinusoidal potential as input signal and current as output signal in low frequency range; d) mass transport due to electrochemical surface reactions as slow process; e) time shift Δt and amplitude of sinusoidal potential $\hat{E}$ and current $\hat{J}$; f) Nyquist diagram with an charge transfer (R_{ct}) and an mass transport (R_{mt}) semi-circle. Partly based on [4, 53].

the semi-circle reflects the resistance R of the process. However, if semi-circles of two processes are overlapping, the analysis is getting more complex. From the distance between the real part of the impedance at the highest frequency $Z_R(f_{max})$ and the ordinate, the ohmic resistance of the system can be determined [62].

3.6 Concluding remarks

In this chapter, fundamentals for developing a dynamic macroscopic ODC model were discussed and a selection of dynamic analysis methods suitable for ODCs was presented. In conclusion, following points must be considered for developing and analyzing a dynamic macroscopic model of an ODC:

- In order to simulate the measured high current densities of ODCs, a TFFA approach seems to be suitable.
- The Henry-constant for oxygen solution and the vapor pressure of water for water evaporation have be set in dependency of the ion and water concentration in the liquid phase.
- The ORR can be described using lumped kinetics and the Tafel equation.
- Due to the multi component diffusion and the high ion concentration in the electrolyte, liquid mass transport have to be calculated by Maxwell-Stefan diffusion.
- Migration can be neglected.
- Due to the phase transitions, a Stefan-flow has to be considered in the liquid and the gas phase.
- As already known from previous studies, the reaction rate is strongly dependent on local concentration gradients [32]. Thus, the model has to be discretized.
- The interaction of the different processes can be analyzed by their characteristic time constants. While the potential steps cover a high range of time constants, the EIS is high-resoluting tool especially for fast processes.

Based on the outcome of this chapter, in chapter 4 a basic model will be developed which depicts all relevant processes. In consideration of (intermediate) results, the model will be extended systematically in chapters 5 and 6. Each development stage of the model is customized to answer specific research questions:

Chapter 4 : Basic model

By means of the basis model, the time constants of the processes of an ODC will be determined. It will be shown that the current density is limited by the availability of dissolved oxygen in the liquid electrolyte and that this is closely linked to ion and water mass transport.

Chapter 5 : Expanded liquid transport model

The model focuses further on mass transport in the liquid phase. The significance of water and ion mass transport will be evaluated in more detail and how these govern the dynamics and performance of ODCs.

Chapter 6 : Inhomogeneous model

The inhomogeneous model enables to consider the distribution of the electrolyte and thus the distribution of the gas-liquid interface within the ODC. It will be shown that only certain areas in the ODC are electrochemically active, although the three-phase boundary extends inhomogeneously over the entire ODC thickness.

The characteristics, implemented processes and assumptions of each model are given in the corresponding chapters. An overview about the differences between the models is given in table 3.1.

Table 3.1: List of properties varying in the different model expansion stages.

Property / Process	Basic model	Expanded liquid mass transport model	Inhomogeneous model
Activity of O_2	ideal	non-ideal	non-ideal
Activity of NaOH	not included	non-ideal	non-ideal
Electrolyte distribution	homogeneous	homogeneous	inhomogeneous
Reaction kinetics	1-step kinetics based on gross reaction	1-step kinetics based on gross reaction	1-step kinetics based on rate determining step
Gradient for diffusion in liquid phase	concentration	concentration	chemical potential

4 Processes and Their Limitations in Oxygen Depolarized Cathodes[6]

In this chapter, the basic dynamic three-phase model of ODCs is introduced and parameterized by using literature data, in order to analyze the processes and limitations within ODCs.

4.1 Introduction

To explore the performance limiting processes of ODCs and to lay the foundation for a systematic improvement, in this chapter a dynamic one-dimensional three-phase model of porous ODCs for the ORR is introduced. To the best of my knowledge, it is the first dynamic model of an ODC and the first ODC model in general, which takes the phase equilibrium dependent on the water and ion concentration in the liquid electrolyte consequently into consideration. These attributes are e.g. beneficial for estimating the location of the gas-liquid interface. In contrast to the stationary models, e.g. the state-at-the-art thin-film flooded agglomerate model by Pinnow et al. [32], time constants can be simulated for a deeper analysis of the limiting factors and their interaction.
In this chapter, polarization curves from literature are used to parameterize the model. Next, dynamic simulations are evaluated to determine the relevant time constants and the limiting processes of the system. Finally, the evaluation is completed with a sensitivity study to show the impact of model parameters on the overall performance of the ODC. Through this procedure, I will show that the mass transport of water hydroxide ions in the liquid electrolyte is determined as the overall restricting factor.

[5] Parts of this chapter have been published in Röhe et al., ChemSusChem, 2019 [1].

4.2 Basic model: Model concept and assumptions

Under operating conditions, the ODC is impinged by oxygen from the one side and by the liquid electrolyte from the other side (cf. fig 4.1)). Due to the combination of the hydrophilic silver catalyst and the hydrophobic PTFE, only a part of the ODC gets flooded so that three different segments within the ODC are expected: The section facing the oxygen inlet contains a pure gas phase, while the section facing the electrolyte is electrolyte flooded. In between the interface of both phases, the so called three phase area, is located. Processes occurring directly at the interface are the evaporation of water and the dissolution of oxygen into the electrolyte. Due to the poor solubility of oxygen into the electrolyte it is estimated that the ORR (eq. (1.1)) takes place in the first few nanometers of the electrolyte [32, 34]. In addition to the transport of the oxygen into the reaction zone, the provision of water and the removal of hydroxide ions are mass transport processes.

To transfer the complex behavior of ODCs to a suitable dynamic model, the following assumptions are made:

1. No pressure drop occurs within the electrode.
2. The temperature in the electrode is constant and homogeneous.
3. Complete ORR (eq. (1.1)) is assumed. The possible incomplete ORR with hydrogen peroxide as final product is not considered, due the high catalytic activity of silver for reducing hydrogen peroxide [31]. The reaction is modeled as lumped one-step kinetics. So it depends on the availability of all reactants.
4. Due to the high overpotential, the anodic current is negligible, hence the kinetic can be described with the Tafel-equation.
5. The location of the gas-liquid interface is stationary over time.
6. Within the electrolyte electroneutrality is assumed.
7. Since the reaction zone in the electrode is very thin and the catalyst as well as the electrolyte are highly conductive, no internal voltage distribution is considered.
8. The high concentration of NaOH results in a low viscosity of electrolyte, thus a liquid diffusion layer between the ODC and the electrolyte bulk has to be considered, even if the electrolyte bulk itself is ideally stirred.

9. A gaseous diffusion layer is not relevant. As will be shown later, the diffusion length in the gas phase does not contribute significantly to the total mass transport resistance and would not affect the ODC performance.

10. In all liquid sections, the multi component diffusion is described by Maxwell-Stefan diffusion. Because the concentration of hydroxide ions and water is each about 10^6 times higher than the oxygen concentration, the influence of the oxygen diffusion on the water and hydroxide diffusion is negligible.

11. The change of the electrolyte concentration due to the ORR causes a change of electrolyte density (cf. fig. B.5). With the further assumption of locally stationary phases, an overlaying convective flux is included, to consider the change of electrolyte volume. To calculate this flux, the simplification $c_{total} = c_{\mathrm{total}}(z=0, t=0)$ is used (cf. fig. B.1). The error is less than 0.8 %, a detailed error estimation is given in the appendix. Further, convection caused by the water evaporation and the oxygen dissolution is implemented in both phases.

12. The concentration of Na^+ is calculated from the equation for electroneutrality. To consider its influence on the mass transport, the binary diffusion coefficient $Đ^{\mathrm{liq}}_{\mathrm{H_2O,NaOH}}$ of NaOH vs. H_2O within the liquid electrolyte is calculated with the concentrated solution theory, which takes OH^- as well as Na^+ into account [60].

13. In the ODC, the gradients in through-plane direction (z-direction) are decisive. A homogeneous distribution of the phases and their components in x- and y-direction in-plane is assumed.

The one-dimensional three-phase model is divided into four spatially resolved sections. As shown in figure 4.1, three of them are located within the porous ODC: A pure gas phase, a thin-film and a flooded porous structure, named flooded agglomerates in the following. The thin-film represents the interface of liquid and gas phase within the electrode. The fourth section is a liquid boundary layer beyond the electrode. For the thin-film, the flooded agglomerates and the liquid diffusion layer, hereafter the collective term liquid phase is used.
The gas phase contains exclusively gases, namely oxygen and water vapor. While the oxygen is provided by the gas bulk phase on the left, the water evaporates from the liquid phase at the right hand side. Additionally oxygen is dissolved in the thin-film and transported into the flooded agglomerates where the ORR occurs. The thin-film itself is an exclusive transport layer, where no reaction takes place. The upstream liquid boundary layer is included, since a significant diffusion resistance is expected [34]. Since a homogeneous distribution of the gas-liquid interface and of the concentration in x- and y-direction is assumed, the specific surface area of the interface

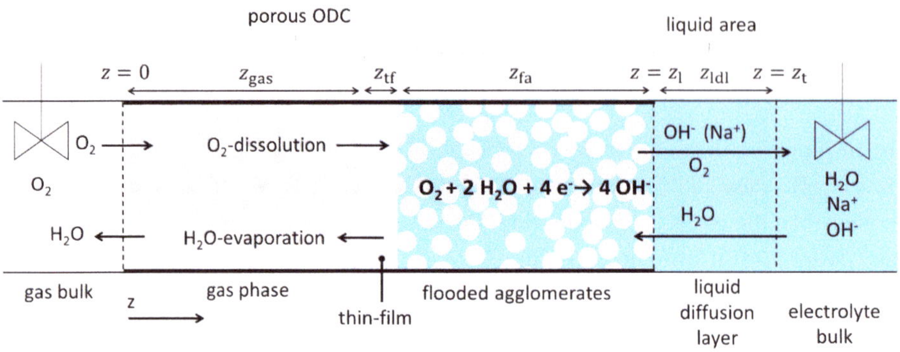

Figure 4.1: Schematic view of the model including reaction and mass transport.

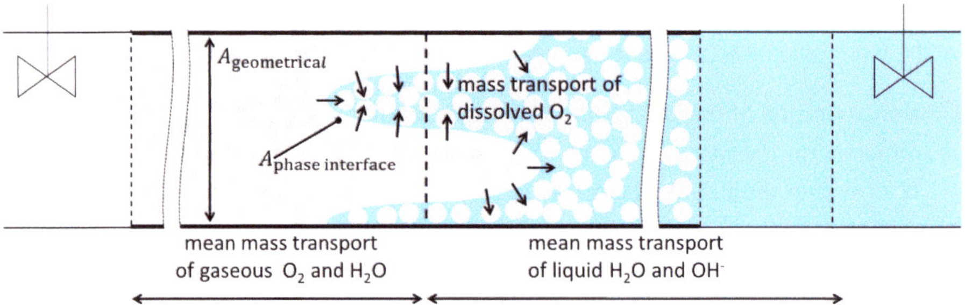

Figure 4.2: Mass transport processes at the gas-liquid interface.

$S_{\text{phase,interface}}$ is defined, which represents the phase interface area per geometric cross sectional area of the electrode:

$$S_{\text{phase,interface}} = \frac{A_{\text{phase,interface}}}{A_{\text{geometric}}} \tag{4.1}$$

This parameter enables the one-dimensional model approach to meet the characteristic of a large gas-liquid-interface within the ODC, as displayed in figure 4.2. Since a steep oxygen gradient within a few nanometers distance from the gas-liquid interface is assumed [34], the meaningful area for dissolution and subsequent mass transport of oxygen is defined to be the specific interface area (cf. fig. 4.2). The thin-film represents the liquid side of the gas-liquid interface, hence the interface area is also relevant for the transport of water and hydroxide ions through the thin-film. In contrast, the area for transport in the gas phase and for the transport of water and the hydroxide ions in the flooded agglomerates is given by the geometric cross section.

4.3 Model equations

4.3.1 Electrochemical reaction kinetics, double layer capacitance and internal resistance

The Faradaic current density j_F is modeled by using Farady's law (4.2) and the Tafel-equation (4.3). All given current densities are referenced to the geometric cross section:

$$j_F = -\mathrm{F} \cdot 4 \cdot r \tag{4.2}$$

Where the reaction rate r is calculated by

$$r = k_0 \cdot a_{\mathrm{O_2}} \cdot a_{\mathrm{H_2O}}^2 \cdot \exp\left(-\frac{(1-\alpha)\cdot \mathrm{F} \cdot \eta_\mathrm{reaction}}{\mathrm{R} \cdot T}\right) \tag{4.3}$$

Where k_0 is the kinetic reaction rate constant, a_i are the activities of the species i, η_reaction is the overpotential, α is the anodic symmetry factor and T is the temperature.
The concentration and temperature dependent activity $a_{\mathrm{H_2O}}$ is based on empirical studies of Balej [66], while for the oxygen activity $a_{\mathrm{O_2}}$ an approximately ideal behavior is assumed:

$$a_{\mathrm{O_2}} = f_{\mathrm{O_2}} \cdot \frac{c_{\mathrm{O_2}}}{c_{\mathrm{O_2}}^{\Theta}} \tag{4.4}$$

Hence $a_{\mathrm{O_2}}$ is calculated with the fugacity coefficient $f_{\mathrm{O_2}} = 1$, the standard concentration $c_{\mathrm{O_2}}^{\Theta} = 1\,\mathrm{kmol\,m^{-3}}$ and the concentration of the dissolved oxygen in the electrolyte $c_{\mathrm{O_2}}$.
The overpotential of the ORR η_reaction is defined as the difference between the open cell voltage E_0 and the potential in the reaction zone E_int:

$$\eta_\mathrm{reaction} = E_\mathrm{int} - E_0 \tag{4.5}$$

Ohmic potential losses can be calculated according to Ohm's law as the difference between the external applied and the potential in the reaction zone:

$$E_\mathrm{ext} - E_\mathrm{int} = R_\mathrm{specific} \cdot j \tag{4.6}$$

Here E_int is the potential in the reaction zone, which equals the external applied potential E_ext under consideration of the specific Ohmic resistance R_specific. The current density j is the sum of the Faradaic and the capacitive current density $j = j_\mathrm{F} + j_\mathrm{dl}$.

Assuming a constant double layer capacitance C_{dl}, the charge balance in the reaction zone is given, as:

$$C_{\mathrm{dl}}\frac{\mathrm{d}E_{\mathrm{int}}}{\mathrm{d}t} = j - \mathrm{F}\cdot 4\cdot r \tag{4.7}$$

4.3.2 Mass transport in the gas phase and processes at the gas-liquid interface

For modelling the mass transport in the gas phase, the evaporation of water and the dissolution of oxygen at the gas-liquid interface have to be taken into account as major effects. Since the absolute values of these two fluxes are typically unequal, in addition to diffusion an overlaying convection has to be taken into account. This process is described by the bilateral non-equimolar diffusion, characterized by Fick's diffusion with an overlaying convective Stefan-flow [54]. The resulting species balance for $i \in \{H_2O, O_2\}$ in the gas phase is:

$$\frac{\partial p_i}{\partial t}\cdot\varepsilon = \frac{\partial^2 p_i}{\partial z^2}\cdot\mathscr{D}_{i,j}^{\mathrm{gas,eff}} + \frac{\partial p_i}{\partial z}\cdot v_{\mathrm{gas}} \tag{4.8}$$

With p_i as the partial pressure of the species i, t as the time, ε as the porosity, z as the location in z-direction, $\mathscr{D}_{i,j}^{\mathrm{gas,eff}}$ as the effective binary diffusion coefficient and v_{gas} as the velocitiy of the convective flow. $\mathscr{D}_{i,j}^{\mathrm{gas,eff}}$ is calculated from the free diffusion coefficient $\mathscr{D}_{i,j}^{\mathrm{gas}}$, porosity ε and tortuosity τ:

$$\mathscr{D}_{i,j}^{\mathrm{gas,eff}} = \mathscr{D}_{i,j}^{\mathrm{gas}}\cdot\frac{\varepsilon}{\tau} \tag{4.9}$$

The tortuosity is approximated with the Bruggemann relation $\tau = \varepsilon^{-0.5}$ [67].
For calculating the Stefan-flow, constant pressure p_{total} is assumed. The only sinks and sources for both components are dissolution and evaporation at the gas-liquid interface $z = z_{\mathrm{gas}}$ and the further inclusion of the ideal gas law results in the same convective velocity over the whole gas phase:

$$\frac{\mathrm{d}v_{gas}}{\mathrm{d}z} = 0 \tag{4.10}$$

Thus the velocity at every point in the gas phase exclusively depends on the mass transfer over the gas-liquid interface:

$$\dot{N}_{\mathrm{total}}^{\mathrm{convection}} = \dot{N}_{\mathrm{H_2O}}^{\mathrm{evaporation}} - \dot{N}_{\mathrm{O_2}}^{\mathrm{dissolution}} \tag{4.11}$$

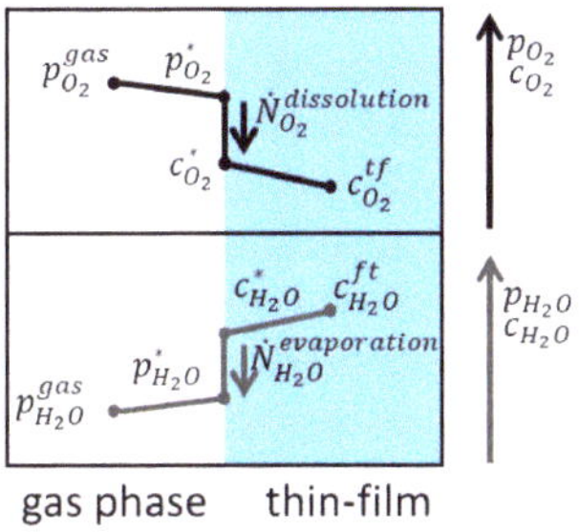

Figure 4.3: Illustration of model on water evaporation and oxygen dissolution immediately at the gas-liquid interface. While the oxygen is solved from the gas phase into the liquid electrolyte (top), the water evaporates from the liquid electrolyte into the gas phase (bottom).

The resulting convective mass flow $\dot{N}_{\text{total}}^{\text{convection}}$ can also be described by the velocity of the flow in gaseous phase:

$$\dot{N}_{\text{total}}^{\text{convection}} = \frac{p_{\text{total}}}{\mathrm{R} \cdot T} \cdot v_{\text{gas}} \tag{4.12}$$

The evaporation of water is described by Raoult's law, where the partial pressure of water $p_{\mathrm{H_2O}}^{*}$ directly at the gas side of the interface equals the vapor pressure $p_{\mathrm{H_2O}}^{\text{vap}}$, which is a function of the concentration of sodium hydroxide c_{NaOH}^{*} and of water $c_{\mathrm{H_2O}}^{*}$ immediately at the interface. The vapor pressure dependence $p_{\mathrm{H_2O}}^{\text{vap}}(c_{\mathrm{NaOH}}^{*}, c_{\mathrm{H_2O}}^{*}, T)$ is taken from by Hirschberg. [51] c_{NaOH}^{*} and $c_{\mathrm{H_2O}}^{*}$ are calculated by considering the mass conservation during the evaporation process with the use of a steady state mass balance (see fig. 4.3):

$$\dot{N}_{\mathrm{H_2O}}^{\text{TF,out}} = \dot{N}_{\mathrm{H_2O}}^{\text{evaporation}} = \dot{N}_{\mathrm{H_2O}}^{\text{Gas,in}} \tag{4.13}$$

Here, $\dot{N}_{\mathrm{H_2O}}^{\text{evaporation}}$ is the mass flow of evaporating water, which equals the sink of water in the liquid phase $\dot{N}_{\mathrm{H_2O}}^{\text{TF,out}}$ as well as the source of water in the gas phase $\dot{N}_{\mathrm{H_2O}}^{\text{Gas,in}}$. By solving the given equations, a dynamic implementation of Raoult's law can be obtained. A detailed description of this procedure can be found in Schröder et al. [12]. For technical implementation, this approach is used to interpolate the concentrations and in particular, the partial pressures from the center of adjacent gas volume element and thin-film volume element at the gas-liquid interface to the common border between these volume elements (cf. fig 4.3).

A similar method is implemented for the dissolution of oxygen in the electrolyte, where the flux leaving the gas phase $\dot{N}_{\mathrm{O_2}}^{\text{Gas,out}}$ is equal to the flux entering the liquid electrolyte $\dot{N}_{\mathrm{O_2}}^{\text{TF,in}}$:

$$\dot{N}_{\mathrm{O_2}}^{\text{Gas,out}} = \dot{N}_{\mathrm{O_2}}^{\text{dissolution}} = \dot{N}_{\mathrm{O_2}}^{\text{TF,in}} \tag{4.14}$$

Here Henry's law comes into effect to determine the oxygen concentration $c^*_{O_2}$ directly at the gas-liquid interface depending on the oxygen partial pressure $p^*_{O_2}$, where the inverse Henry constant for oxygen dissolution H_{O_2} is a function of c^*_{NaOH} and $c^*_{H_2O}$ as well [22]:

$$c^*_{O_2} = \frac{p^*_{O_2}}{H_{O_2}(c^*_{NaOH}, c^*_{H_2O})} \tag{4.15}$$

The concentration dependent functions of the water vapor pressure $p^{vap}_{H_2O}(c_{NaOH}, c_{H_2O})$ (fig. B.3) and the Henry constant for oxygen dissolution $H_{O_2}(c_{NaOH}, c_{H_2O})$ (fig. B.2) are shown in the appendix B.

4.3.3 Mass transport in the thin-film

The mass transport in the thin-film is characterized by a diffusion process with an overlaying convection of all species due to the evaporation. This results from the assumption of a non-moving interface and a constant thin-film volume. Since there are three species i in the thin-film, $i \in \{OH^-, H_2O, O_2\}$, for the diffusion process the Maxwell-Stefan diffusion with the concentration gradient as driving force is taken into account.

$$\frac{\partial c_i}{\partial t} = \sum_{\substack{j=1 \\ i \neq j}}^{n} \left(\frac{\partial^2 c_i}{\partial z^2} \frac{c_j}{c_{total}} - \frac{\partial^2 c_j}{\partial z^2} \frac{c_i}{c_{total}} \right) \cdot Đ^{liq}_{i,j} + \frac{\partial c_i}{\partial z} \cdot v_{liq,evaporation} \tag{4.16}$$

Again, the oxygen concentration is very low, thus no reasonable effect from the oxygen diffusion to the diffusion of the water and hydroxide ions is expected, which leads to $j \in \{OH^-, H_2O\}$. As in the gas phase, the liquid evaporation velocity $v_{liq,evaporation}$ is constant throughout the thin-film.

$$\frac{dv_{liq,evaporation}}{dz} = 0 \tag{4.17}$$

Thus, the velocity of the convective flow is directly proportional to the evaporation rate.

$$v_{liq,evaporation} = \frac{\dot{N}^{gas,\ in}_{H_2O}}{c^*_{H_2O}} \tag{4.18}$$

4.3.4 Mass transport in the flooded agglomerates

The mass transport within the flooded agglomerates is also described by the Maxwell-Stefan diffusion with overlaying convection v_{liq}. The effective diffusion coefficient takes the porous structure into account and is calculated by equation (4.9). Moreover the sinks and sources due to the ORR have to be modeled. This yields the following species balance:

$$\frac{\partial c_i}{\partial t} \cdot \varepsilon = \sum_{\substack{j=1 \\ i \neq j}}^{n} \left(\frac{\partial^2 c_i}{\partial z^2} \frac{c_j}{c_{\text{total}}} - \frac{\partial^2 c_j}{\partial z^2} \frac{c_i}{c_{\text{total}}} \right) \cdot Đ_{i,j}^{\text{liq, eff}} + \frac{\partial c_i}{\partial z} \cdot v_{\text{liq}} - r \tag{4.19}$$

With $i \in \{OH^-, H_2O, O_2\}$, $j \in \{OH^-, H_2O\}$.

In addition to the above described convection caused by the evaporation, there is a further effect due to the reaction. The ORR causes a local change of water and hydroxide concentration, which consequently leads to a change in liquid density (cf. fig. B.5). Although the molecules of water and hydroxide per volume change depending on the reaction, the total concentration c_{total} stays almost totally constant (cf. fig. B.1). Consequently, the additional Stefan-flow based on the density change between the reaction area and the electrolyte bulk phase can be described by the removal of the additional produced ions:

$$v_{\text{liq,react}}(z) = \left(|\nu_{H_2O}^{\text{ORR}}| - |\nu_{OH^-}^{\text{ORR}}|\right) \int_0^z \frac{r(z)}{c_{\text{total}}} \mathrm{d}z \tag{4.20}$$

where

$$\frac{\mathrm{d}v_{\text{liq, react}}}{\mathrm{d}z} \neq 0 \tag{4.21}$$

Here, r is the reaction rate, $\nu_{H_2O}^{\text{ORR}}$ the stoichiometric factor of water and $\nu_{OH^-}^{\text{ORR}}$ stoichiometric factor of hydroxide ions in the ORR.

Finally the local total velocity v_{liq} for each location z can be calculated by the velocity of the evaporation $v_{\text{liq,evaporation}}$ and the location-dependent velocity of reaction $v_{\text{liq, react}}$:

$$v_{\text{liq}}(z) = v_{\text{liq,evaporation}} + v_{\text{liq, react}}(z) \tag{4.22}$$

4.3.5 Mass transport in the liquid diffusion layer

The mass transport in the liquid diffusion layer is described by the same equations as the mass transport within the flooded agglomerates. In this layer, free diffusion occurs, so the free binary diffusion coefficient is used:

$$\frac{\partial c_i}{\partial t} = \sum_{\substack{j=1 \\ i \neq j}}^{n} \left(\frac{\partial^2 c_i}{\partial z^2} \frac{c_j}{c_{\text{total}}} - \frac{\partial^2 c_j}{\partial z^2} \frac{c_i}{c_{\text{total}}} \right) \cdot Đ_{i,j}^{\text{liq}} + \frac{\partial c_i}{\partial z} \cdot v_{\text{ldl}} \tag{4.23}$$

With $i \in \{OH^-, H_2O, O_2\}$, $j \in \{OH^-, H_2O\}$.
Since there are no sinks and sources in the liquid boundary layer, the convective velocity v_{ldl} is equal to that at $z = z_l$ (cf. fig 4.1) of the flooded agglomerates:

$$v_{ldl} = v_{liq}(z = z_l) \tag{4.24}$$

, where

$$\frac{dv_{ldl}}{dz} = 0 \tag{4.25}$$

4.3.6 Boundary conditions

Due to the high flux of oxygen in the gas bulk phase, the following boundary conditions are defined:

$$p_{O_2}(z = 0) = p_{total} \tag{4.26}$$

$$p_{H_2O}(z = 0) = 0 \tag{4.27}$$

Since the electrolyte bulk phase is assumed to be ideally stirred, oxygen is expected to be zero.

$$c_{O_2}(z = z_t) = 0 \tag{4.28}$$

The amount of water in the electrolyte bulk phase can be calculated as a function of sodium hydroxide concentration [68], which is constant.

$$c_{NaOH}(z = z_t) = c_{NaOH}(z = z_t, t = 0) \tag{4.29}$$

$$c_{H_2O}(z = z_t) = f(c_{NaOH}(z = z_t, t = 0)) \tag{4.30}$$

4.4 Parameterization

The model was implemented in MATLAB. For parameter identification, data by Pinnow et al. was used. The examined ODC was manufactured by a spray method with a content of 97 wt−% silver and wt−% PTFE. The experiments were performed at 80 °C and with a NaOH molality of $m_{NaOH} = 11.25\,\mathrm{mol\,kg^{-1}}$ [32], which under the given conditions is a NaOH concentration of

$c_{\mathrm{NaOH}}(z = z_\mathrm{t}) = 10.1 \times 10^3\,\mathrm{mol\,m^{-3}}$ [51, 68]. All further constant model parameters, set to their respective values from the experiment, are given in table 4.2, and concentration dependent parameters are listed in table 4.2. The literature data were measured from $j = 0$ to $j = 4\,\mathrm{kA\,m^{-2}}$ and simulated above $j = 4\,\mathrm{kA\,m^{-2}}$ by the same authors [32, 39].
Identified model parameter are the kinetic reaction rate constant k_0, the specific interface area $S_{\mathrm{phase,interface}}$, the specific resistance R_{specific} and the length of flooded agglomerates z_{fa}. The manually adjusted parameters which best reproduce the polarization curve documented by Pinnow et al. are given in table 4.3. The corresponding curve in figure 4.4 shows, that the data matches very well, thus the model can reproduce the steady state polarization curve.
The identified specific gas-liquid interface area $S_{\mathrm{phase,interface}}$ is about 2.5 times higher than the interface modeled by Pinnow et al. [32]. This deviation may originate from the here applied strict consideration of the ion concentration dependent equilibrium at the gas-liquid interface. The ion concentration dependency of the phase equilibrium also enables an estimation of the location of the gas-liquid interface: The average value of the distributed gas-liquid interface is located $11.5\,\mu\mathrm{m}$ from the end of the electrode facing the liquid side, so that only a small part of the ODC is flooded with the liquid electrolyte (fig. 4.2). Since the diameter of the pores of comparable ODC is typically less than one micro meter [69] and the PTFE has strong hydrophobic attributes, this value seems to be valid. Present theories about the shape of the gas-liquid interface are e.g. cylindrical electrolyte flooded agglomerates which are surrounded by the gas phase [32] or microscopic gas channels reaching into the flooded electrode part [70]. When assuming a regular cylindrical shape of the catalyst agglomerates in the electrode which are oriented perpendicular to the electrode surface [39], agglomerate diameters of $d_{\mathrm{ag}} = 0.62\,\mu\mathrm{m}$ are obtained. This comparably small diameter may indicate microscopic gas channels into the flooded agglomerates, an non plane surface following the catalyst particles or a moving interface.
As discussed in chapter 4.5.3, all identified parameters have a characteristic influence on the polarization curve, which is indicated by different sensitivities of the overpotential to the parameters at kinetic, ohmic and mass transport governed operation conditions. Nevertheless, the identified parameter set may not be unique. Reliable parameterization would require additional, especially dynamic measurements, which enable a strict separation of slow and fast processes. However, based on the comparison to literature values, I assess the parameter set to be reasonable.

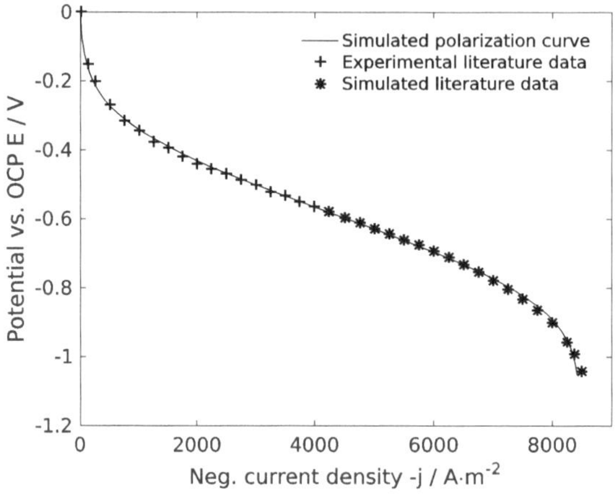

Figure 4.4: Simulated polarization curve in comparison to literature data [32].

4.5 Results and discussion

4.5.1 Steady state analysis

To develop a deeper understanding of the processes within the ODC, the states underlying the simulated polarization curve (cf. fig. 4.4) are analyzed in the following section in detail. The electrode potential is referred to the open cell potential E_0 of the simulated process conditions: $E = E_{\text{ext}} - E_0$. As shown in figure 4.5 a), the oxygen concentration depletes strongly within the electrode close to the gas-liquid interface. The drop gets steeper with increasing negative current densities. Even at medium current densities of about $5\,\text{kA}\,\text{m}^{-2}$, dissolved oxygen can be found only in the first 200 nm of the flooded electrode section. Although only a small fraction of the catalyst is involved in the reaction, the kinetic losses are comparably low, which indicates a high catalytic activity of the silver for the ORR in alkaline media. At higher current densities, only the region directly behind the interface stays electrochemically active, which supports the findings of [32, 34]. Here, two major effects are identified, leading to the transport limitation of oxygen: Firstly, for increasing current, more oxygen is consumed by the ORR, which is indicated by the steeper gradient of oxygen concentration in the thin-film and the flooded agglomerates. Secondly, the higher reaction rate at higher currents causes an enrichment of ion concentration of OH^- as the reaction product of the ORR and an accompanying decline in water concentration as an educt of the ORR (cf. figure 4.5 b). Since the Henry constant is a function of ion concentration (cf. fig. B.2), the solubility of oxygen decreases at higher current densities and limits the amount of dissolved oxygen in the liquid electrolyte. This effect can be recognized by the decreasing oxygen concentration at the gas-liquid interface. The high sensitivity of oxygen solubility to the

Table 4.1: List of constant parameter and operation conditions.

Name	Symbol	Unit	Value
Electrode geometry			
Electrode thickness [32]	$z_{\text{electrode}}$	m	300×10^{-6}
Thin-film thickness [32]	z_{tf}	m	60×10^{-9}
Liquid diffusion layer thickness[a]	z_{ldl}	m	50×10^{-6}
Porosity [32]	ε	–	0.4
Tortuosity [67]	τ	–	1.58
Operation conditions			
Pressure in gas chamber[b] [32]	$p = p_{O_2}$	Pa	1×10^5
NaOH concentration in electrolyte bulk[b] [32]	$c_{\text{NaOH}}(z = z_l)$	mol m^{-3}	10.1×10^3
H_2O concentration in electrolyte bulk [51]	$c_{H_2O}(z = z_l)$	mol m^{-3}	49.7×10^3
Temperature [32][b]	T	K	353.15
Electrochemical and kinetic data			
Open circuit potential [71]	E_0	V	0.216
Symmetry factor[c] [24]	α	–	0.15
Double layer capacitance [72]	C_{dl}	F m^{-2}	50
Diffusion coefficients			
Binary gas diffusion coefficient [73]	$\mathscr{D}^{\text{gas}}_{O_2,H_2O}$	$\text{m}^2\,\text{s}^{-1}$	2.98×10^{-5}
Binary liquid diffusion coefficient[d] [73, 74]	$Đ^{\text{liq}}_{O_2,H_2O}$	$\text{m}^2\,\text{s}^{-1}$	2.12×10^{-9}
Binary liquid diffusion coefficient[d] [73, 74]	$Đ^{\text{liq}}_{O_2,\text{NaOH}}$	$\text{m}^2\,\text{s}^{-1}$	3.16×10^{-9}
Binary liquid diffusion coefficient[e] [60]	$Đ^{\text{liq}}_{H_2O,\text{NaOH}}$	$\text{m}^2\,\text{s}^{-1}$	1.19×10^{-9}

[a] Own estimation. [b] Experimental operation conditions from [32]. [c] Measured with 6.5 M NaOH.
[d] Calculated with modified Wilke and Chang equation, $\Theta = 3.9$ and $a = 0.5$ chosen, to fit the in [74] given experimental data best. [e] No NaOH data available, modeled for KOH with NaOH viscosity from [68], as seen in [32].

Table 4.2: Electrolyte concentration depended variables.

Name	Symbol	Unit
Inverse Henry constant[a] [22]	H_{O_2}	$\text{Pa m}^3\,\text{mol}^{-1}$
Activity of H_2O[b] [75]	a_{H_2O}	–
Density of electrolyte [68]	ρ	kg m^{-3}
Water vapor pressure [51]	$p^{\text{vap}}_{H_2O}$	Pa

Dependencies are shown in fig. B.1-B.5
[a] Extrapolated from $c_{\text{NaOH}} \leq 6\,\text{M}$, but as shown in [27] extrapolated values fit very well to experiments $c_{\text{NaOH}} \leq 12\,\text{M}$. [b] Extrapolated data from 343.15 K to 353.15 K.

ion concentration also explains the in previous studies documented mass transport impact even at low current densities of about $0.1\,\text{kA m}^{-2}$ [39]. Water concentration only declines by 14.3 % to $43.5 \times 10^3\,\text{mol m}^{-3}$ at $j = 8.4\,\text{kA m}^{-2}$, whereas the ion concentration increases by 37.3 % to

Table 4.3: Identified parameters.

Name	Symbol	Unit	Value
Reaction kinetic rate constant	k_0	$\mathrm{mol\,s^{-1}m^{-2}}$	165
Specific interface	$S_{\mathrm{phase,interface}}$	$\mathrm{m^2\,m^{-2}}$	134
Overall specific resistance	R_{specific}	$\Omega\,\mathrm{m}^2$	1.9×10^{-5}
Length of flooded agglomerates	z_{fa}	m	11.5×10^{-6}
Length of gas phase [a]	z_{gas}	m	288×10^{-6}

[a] Calculated by $z_{\mathrm{gas}} + z_{\mathrm{tf}} + z_{\mathrm{fa}} = z_{\mathrm{electrode}}$

$16.1 \times 10^3\,\mathrm{mol\,m^{-3}}$. Hence, significant consequences do not result from the lower water concentration being available for the reaction, but the mentioned change of the phase equilibrium. The change of water concentration at the interface of the flooded agglomerates to the liquid diffusion layer shows that not only the mass transport within the ODC is important, but also the mass transport across the liquid diffusion layer.

For analyzing the driving forces of the mass flow in the liquid phase, the fraction of mass flow driven by diffusion $\dot{N}_{\mathrm{i}}^{\mathrm{diffusion}}$ in relation to the total mass flow $\dot{N}_{\mathrm{i}} = \dot{N}_{\mathrm{i}}^{\mathrm{diffusion}} + \dot{N}_{\mathrm{i}}^{\mathrm{convection}}$ of the species i is calculated by $\dot{x}_{\mathrm{i}} = |\dot{N}_{\mathrm{i}}^{\mathrm{diffusion}}|/|\dot{N}_{\mathrm{i}}|$. For all species the mass flow in the liquid phase is dominated by diffusion. As shown in figure 4.6, the oxygen mass transport in particular is almost 100% diffusion driven, which is due to the steep concentration gradient and the low concentration in the flooded agglomerated. Since oxygen depletes with increasing depth into the flooded agglomerates, an evaluation of the oxygen mass transfer is only directly behind the thin-film possible (cf. fig. 4.6). For water and hydroxide ions lower values are obtained: $\dot{x}_{\mathrm{H_2O}} \approx 70\,\%$ and $\dot{x}_{\mathrm{OH^-}} \approx 90\,\%$ respectively. Since the concentration gradient of water and hydroxide ions are practically the same but in opposite direction within the flooded agglomerates, the diffusion mass flow is in the same range for both species as well. In contrast, the convection-based mass flow of each species is proportional to the concentrations, which is higher for the water than for the hydroxide ions. Accordingly, relative to the total mass transfer, the diffusion driven mass transfer for water is the lowest and for oxygen, the highest. The high ratio of convective mass flow of water and hydroxide at low currents close to the thin-film is caused by the evaporation of water.

The concentration dependent equilibrium at the gas-liquid interface also affects the partial pressures and the mass transfer in the gas phase. The vapor pressure of the water decreases by a factor of 2.93 (cf. fig. B.3) and the inverse Henry constant increases by a factor of 5.81 (cf. fig. 4.3) when increasing current to maximum. As a consequence of the changes equilibrium for water evaporation and oxygen dissolution, the velocity of convection decreases from $v_{\mathrm{gas}}(j=0) = 7.18 \times 10^{-5}\,\mathrm{m\,s^{-1}}$ to $v_{\mathrm{gas}}(j=j_{\mathrm{max}}) = 1.78 \times 10^{-5}\,\mathrm{m\,s^{-1}}$. As a result, the partial pressure of water is decreasing with higher current densities.

To get a deeper impression of the outlined findings and to separate the different effects, as well as getting information about their interaction, dynamic results are presented in the next section.

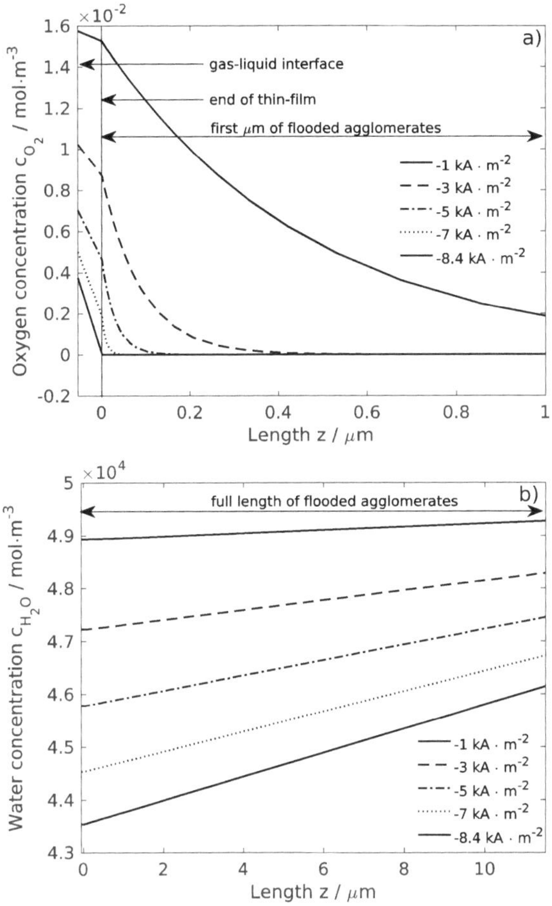

Figure 4.5: Dependence of simulated a) oxygen and b) water concentrations profiles within the liquid phase on current density.

4.5.2 Dynamic analysis

In this section, dynamic simulations with potential steps will be presented, which enable a separation and identification of the interaction of the reaction and the different mass transfer processes. To cover the typical operation current densities of -4 to $-6\,\mathrm{kA\,m^{-2}}$ [20], potential steps from $E=-0.55\,\mathrm{V}$ to $E=-0.65\,\mathrm{V}$ and from $E=-0.65\,\mathrm{V}$ to $E=-0.55\,\mathrm{V}$ vs. OCP have been chosen, which correspond to steady state currents of -3.76 and $-5.30\,\mathrm{kA\,m^{-2}}$. The applied potential E and the current density j as the system response is depicted in figure 4.7. Two significant different time constants can be distinguished: A rapid and strong change of the current within

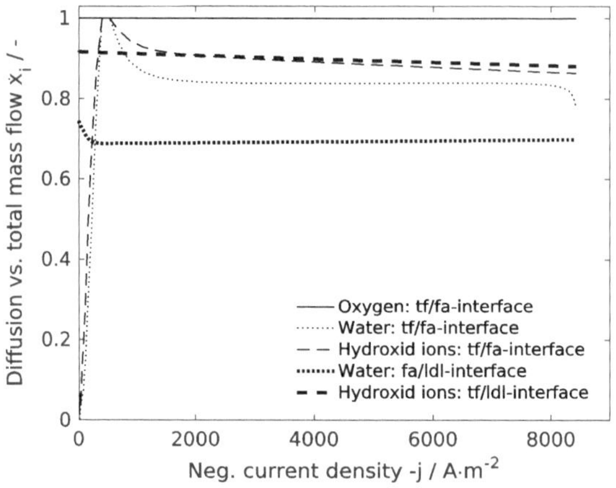

Figure 4.6: Fraction of mass flow driven by diffusion in comparison to the total mass flow in the flooded agglomerats as a function the current density at various locations.

milliseconds and a slow relaxation of smaller amplitude within 5 s. The response signal to step-up and step-down in potential are identical, but with opposite sign. The rapid change observed for the positive as well as for the negative step results from the charging or discharging of the double layer.

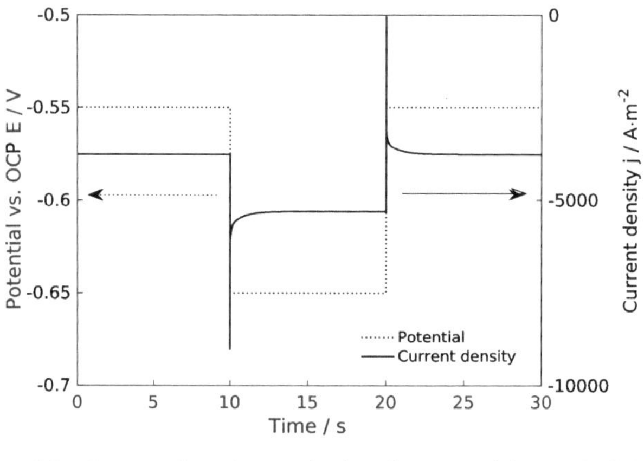

Figure 4.7: Electrode potential and current dependent on the time for a potential step of -0.55 to -0.65 V vs. OCP at $t = 10$ s and back at $t = 20$ s.

The second observed, significantly larger time constant results from the mass transfer within the ODC following the consumption and production of species, but also from the potential sensitive charge transfer. In figure 4.8 a) it can be seen that after the potential step, both water and oxygen concentration need approximately five seconds to reach the new steady state, no overshooting is observed. Since the diffusion pathway of water respectively the hydroxide ions between reaction zone and bulk phase is much longer than the diffusion pathway of oxygen through the thin-film into the reaction area, at least two time constants for the mass transfer were expected. Indeed in figure 4.8 a) one may see that the oxygen concentration change is significantly steeper in the first second, but then slows down and reaches steady state at the same time as water. Comparing both plots in that figure, the concentration of water and oxygen at the boundary between thin-film and flooded agglomerates shows the same relaxation time as the current. This supports the conclusion, that the shift of the phase equilibrium couples both processes and dominates the dynamic response.
To analyze the time behavior of oxygen mass transport, the oxygen partial pressure in the gas phase has to be taken into account (see fig. 4.8 b)). Oxygen partial pressure oddly rises simultaneously with stronger current. Two aspects can be identified: Firstly, although more oxygen is consumed at higher current densities, the partial pressure of oxygen in the gaseous phase is increasing. Secondly the comparison of the dynamics in the gas phase with the dynamics in the liquid phase shows almost the same shape and time constant, suggesting similar rate determining steps. The rise in partial pressure can be attributed to the lowering of water activity due to the production of hydroxide ions and the consumption of water, which decreases the oxygen solubility, but also decreases the vapor pressure of the electrolyte. For deeper analysis, the time constants of the different mass transport processes can be approximated by equation (4.31), under the valid assumption that the mass transfer is dominated by diffusion (see. fig. 4.6).

$$\tau_{\mathrm{mt}} \approx \frac{\Delta z^2}{D} \tag{4.31}$$

Since the oxygen partial pressure in the gaseous phase is increasing with higher current densities, the sinking concentration of oxygen in the liquid phase at high current densities can be attributed to the mass transport through the liquid phase, but not by the gas phase. Inserting the thickness of the thin-film as diffusion length, a time constant of about $\tau_{\mathrm{mt,O_2}} \approx 1.70 \times 10^{-6}\,\mathrm{s}$ is calculated, which explains the significantly steeper fall of oxygen concentration than of water concentration in the first milliseconds after the potential step. For the mass transport of water and hydroxide ions between the reaction zone and the bulk phase, the overlapping time constants for the effective diffusion in the porous structure of the flooded agglomerates $\tau^{\mathrm{fa}}_{\mathrm{mt,H_2O}}$ and the free diffusion in the liquid diffusion layer $\tau^{\mathrm{ldl}}_{\mathrm{mt,H_2O}}$ have to be determined separately. Here, $\tau^{\mathrm{ldl}}_{\mathrm{mt,H_2O}} \approx 2.10\,\mathrm{s}$ and under consideration of the porous structure (cf. eq. (4.19)), $\tau^{\mathrm{fa}}_{\mathrm{mt,H_2O}} \approx 0.176\,\mathrm{s}$ are calculated. Due to the more rapid decreasing of the water concentration within about 1 s after the potential jump on the one hand and to the comparably long relaxation time to a new steady state of about 5 s on the other hand (cf. fig. 4.8 a)), it can be concluded that the observed current density response is dominated by both, the mass transport in the flooded agglomerates and the mass transfer in the liquid diffusion layer.

Combining the given information it can be concluded, that the process governing the dynamic behavior is the slow mass transport of water respectively hydroxide ions through the liquid phase. Thus their interplay with the phase equilibrium dominates the overall performance of the ODC.

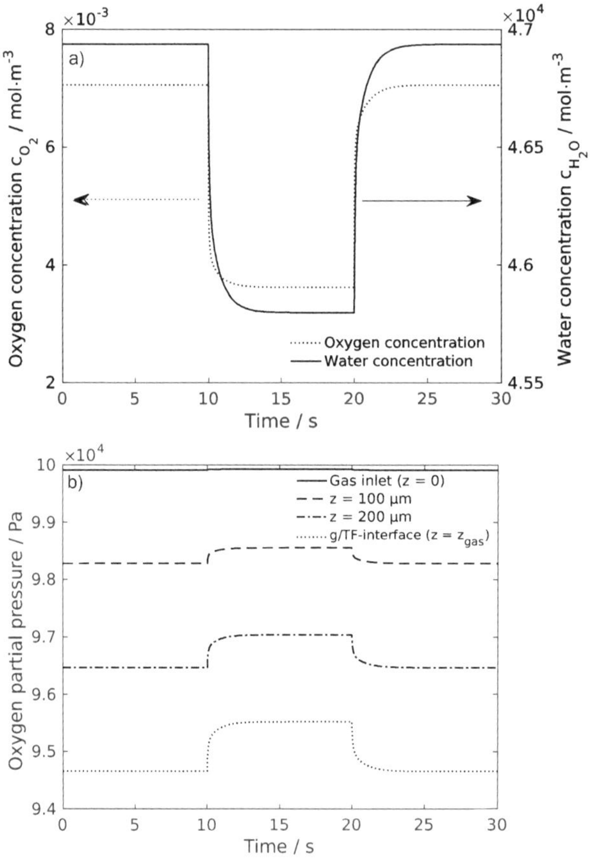

Figure 4.8: a) Concentration responses of oxygen and water at the thin-film-liquid interface for a potential step of −0.55 to −0.65 V vs. OCP at $t = 10$ s and back at $t = 20$ s and b) response of oxygen partial pressure at various locations.

4.5.3 Parameter sensitivities

To analyze the influence of the parameters for the evaluated electrode on the model output systematically, a local sensitivity analysis has been performed. The relative sensitivity s_ς of the electrode potential E_{Ext} to the parameter ς is determined as follows:

$$s_\varsigma = \frac{\frac{E_{ext}-E^\circ_{ext}}{E^\circ_{ext}}}{\frac{\varsigma-\varsigma^\circ}{\varsigma^\circ}} \tag{4.32}$$

Where ς^θ denotes the value at reference point and ς the one for an increased or decreased parameter. Every parameter is varied $\pm$ 5%: $\varsigma = \varsigma^\circ \cdot (1 \pm 0.05)$. The results given in figure 4.9 are the mean of individual results of the positive and the negative modified parameter. A high sensitivity of a parameter indicates a strong impact on the electrode potential, while a sensitivity close to zero indicates a negligible influence [76].

Sensitivity analysis is presented for three characteristic regimes (cf. fig. 4.4): Current densities of $4\,\mathrm{kA\,m^{-2}}$ as the industrial operating point, $0.5\,\mathrm{kA\,m^{-2}}$ as a kinetically governed operating point, and $8\,\mathrm{kA\,m^{-2}}$ as a mass transport governed operating point were analyzed. For the evaluation, it is important to notice that the slope of the polarization curve (cf. fig. 4.4) at $0.5\,\mathrm{kA\,m^{-2}}$ and $8\,\mathrm{kA\,m^{-2}}$ is much higher than the slope at $4\,\mathrm{kA\,m^{-2}}$. Consequently, the parameters may exhibit smaller sensitivities at $4\,\mathrm{kA\,m^{-2}}$ than at the other two points of the polarization curve. The varied parameters are further segmented into three different groups: Kinetic parameters, structural and geometric parameters and operation conditions. The results are given in figure 4.9.

As expected, the reaction rate constant k_0 affects the potential most at low currents. But although the performance at high current densities is dominated by mass transport, the performance shows significant sensitivity to high catalyst activity and a high catalyst surface. The symmetry factor α has a medium-high influence on the ODC over the entire operation range.

To evaluate the sensitivity of the thickness of the different electrode sections individually, the boundary condition $z_{gas} + z_{tf} + z_{fa} = z_{electrode}$ was ignored. The thickness of the liquid phase z_{tf} has the largest influence on the cell potential, while the length of the gas phase z_g is not selective. This supports the prior conclusion of a limitation caused by the mass transport of water and hydroxide ions across the liquid phase, while the mass transport in the gas phase is not limiting. ODC performance may thus be improved by a reduction of the flooded part, especially at higher currents. This could be reached by a more hydrophobic behavior of ODCs, using either smaller pores or a higher amount of the hydrophobic PTFE. But since a lower porosity and tortuosity due to smaller pores as well as a lower reaction rate due to a smaller catalyst surface would lead to a decrease of cell performance, an optimum of the mentioned parameters has to be found. Further, the length of the flooded part z_{fa} can not be arbitrarily reduced, because a certain space for large specific interface is needed. The same holds for the diffusion layer in the liquid electrolyte:

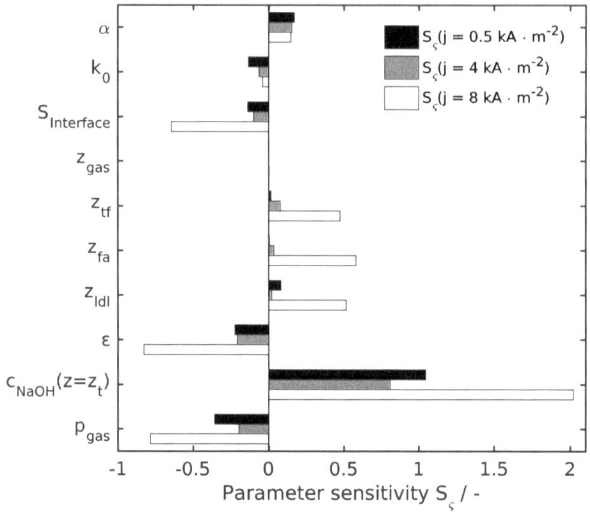

Figure 4.9: Relative sensitivity s_ς of the electrode potential to parameter variation of $\pm$ 5% for three different constant current densities. When increasing c_{NaOH}, the current density of $j = 8\,\mathrm{kA\,m^{-2}}$ could not be reached, thus only results for negative variation are shown.

A thicker liquid diffusion layer has especially at higher current densities a negative impact to operation conditions of the electrode, here an approach to improve the electrode performance is given by an forced convective flow to reduce the diffusion layer thickness z_{ldl}. Moreover, the thin-film acts as an additional diffusion resistance for oxygen dissolved by the gaseous phase [38], thus the high sensitivity of the thin-film thickness at the diffusion limiting current densities is explainable. It has been shown that especially the thin-film thickness dominates the shape of the polarization curve [45]. Finally, the specific interfacial area $S_{\mathrm{phase,interface}}$ between liquid and gas phase is particularly important at high current densities, where oxygen solubility limits the performance.

The large sensitivity to the electrolyte concentration $c_{\mathrm{NaOH}}(z = z_t)$ emphasizes the previously discussed solubility limitation by the ion concentration at the gas-liquid-interface. Previous studies have shown that the highest current densities for the ORR could be achieved with electrolyte concentrations of 5 M [70]. I suggest that process conditions with high concentrations over 5 M leads to a mass transport limitation. But as the hydroxide ions are the desired product of the ORR in advanced chlor-alkali electrolysis, the only way for a higher efficiency of the process is a faster removal of ions. The potential is sensitive to oxygen partial pressure. The decrease of the electrode potential with increasing pressure p_{gas} results from lower mass transfer limitations of the oxygen due to the higher amount of dissolved oxygen into the electrolyte. Although a better performance with higher pressure is documented [70], it has to be mentioned that the pressure of the oxygen can only be increased in a defined range to avoid a break through of the gas to the electrolyte side [9]. Also notable is the sensitivity for the pressure even at low current densities, which meets the results of former studies, that the mass transport influences the ODC losses [39] even at kinetically controlled current densities.

4.6 Conclusions

The first dynamic three-phase model of a porous ODC for advanced chlor-alkali electrolysis has been presented. Suitable model parameters were partly obtained from literature and partly identified from published polarization curves.
It was shown that at industrial operation conditions, the ORR only takes place close to the gas-liquid interface, because of depletion of the dissolved oxygen. The consumption of water and the production of hydroxide ions leads to a significant change of the phase equilibrium with current density, which causes a lower solubility of the oxygen into the liquid electrolyte. This effect intensifies at higher reaction rates and causes a total depletion of oxygen and thus the limiting current. Water is sufficiently available for the reaction and does not limit the maximum current density in a direct way. A deeper dynamic evaluation of this behavior pointed out that the removal of ions from the reaction area is the slowest process. It has a significant impact not only on stationary performance but also on the full dynamic behavior of all concentration profiles and thus of the dynamic response. In this context, it was also shown that only a few micrometers of the ODC are flooded with the liquid electrolyte under industrial working conditions. Hence, dynamic analysis is a highly potent method to reveal the interaction of processes in ODCs and to determine its state. Dynamic measurements should be conducted for reliable, unique parameterization and for further validation of the model.
A sensitivity analysis of the kinetic, geometric and process parameters indicated that an improvement of the ODC performance can be affected most with a fast mass transport in the liquid phase, e.g. via a smaller level of flooding within the electrode or a smaller liquid diffusion layer.

5 In-depth Analysis of Water Mass Transport and Water Activity[7]

In this chapter, the previously presented basic model is enlarged and used to analyze dynamic and steady-state measurements, in order to determine the role of water activity and water mass transport on ODC performance and dynamics.

5.1 Introduction

In the previous model-based study in chapter 4, I analyzed the limiting processes of ODC. I have shown that the consumption of water and the production of hydroxide ions induced by the electrochemical ORR lead to an enrichment of hydroxide ions at the gas-liquid interface within the liquid electrolyte. This causes a reduction in availability of water, a change of the gas-liquid equilibrium and in consequence a substantial decrease of oxygen solubility. Based on these results I hypothesized that a fast mass transport of water and hydroxide ions across the porous electrode and the electrolyte diffusion layer outside the electrode are key drivers for a high ODC performance. Those findings are supported by experiments of Clausmeyer et al. who achieved higher current densities in a 5 M compared to a 10 M NaOH electrolyte utilizing single nanoparticle studies [70], and by Botz et al. who measured an accumulation of hydroxide ions in the diffusion layer in close proximity to the ODC surface at higher current densities [34]. Furthermore, Zhang et al. presented a decreasing oxygen diffusivity in line with an increasing NaOH concentration and a decreasing water activity as a further limiting effect in ODCs [27].
The aforementioned studies demonstrate that the electrolyte has a significant influence on the ODC performance. Under careful consideration of the non-ideal behavior of water as a solvent, the impact on the electrocatalytic properties of an ODC may be explained: In highly concentrated aqueous electrolytes, the activity of water, and its thermodynamics [77], cannot be described by dilute solution theory. Instead, the activity coefficient strongly depends on the ion concentration [75]. The underlying mechanism of a decrease in water activity at increasing NaOH concentration is the formation of solvation shells, in which water molecules are bound to ions. In alkaline

7 Parts of this chapter have been published in Röhe et al., ChemElectroChem, 2019 [2].

media, for instance, the first layer of the hydration shell of a hydroxide ion consists on average of four water molecules, whereas the coordination number decreases to about three at higher ion concentrations [78]. A decreased water activity goes, apart from the lower solubility of oxygen [50], along with a change in mass transport [79], a lower vapor pressure of the water [51] and slower reaction rate [27]. To completely elucidate the influence of the water activity on the ODC performance, a thorough analysis is required.

In this chapter I will confirm the hypothesis that water activity and water mass transport represents a crucial factor in governing the ODC performance. For this purpose, the model was extended in order to resolve the influence of the water and hydroxide activity more precisely. Stationary and dynamic measurements in two different electrochemical setups were conducted by cooperation partners[8] and analyzed with the extended one-dimensional dynamic three-phase model measurements. Both setups had comparable conditions concerning the oxygen gas supply, yet they differed in the respective electrolyte conditions: The first experimental setup was operated under convective electrolyte flow, whilst a stagnant electrolyte was present in the second one. Additionally, in the latter setup, the analysis was assisted by scanning electrochemical microscopy (SECM), which allows the determination of hydroxide and water activities at a distance of about 1 μm above the ODC surface. The differences in the dynamic behavior and steady state performance of both experimental cells are analyzed using the dynamic model. By comparing the two cells, the influence of liquid phase mass transport and water activity in the liquid phase was assessed model-based. It is shown, that both systems differ significantly in their performance and dynamics. The model-based analysis of dynamic measurements and the SECM measurements demonstrated that the performance and the dynamics of ODCs are highly dependent on the mass transport of water and hydroxide ions between the bulk phase and the electrode.

5.2 Experimental

Two different electrochemical cells, as schematically depicted in fig. 5.1, were utilized and both were operated using identical ODCs (see table 5.1). The measurements have been performed by the Institute of Chemical and Electrochemical Process Engineering, Technische Universität Clausthal, Germany and Analytical Chemistry — Center for Electrochemical Sciences, Ruhr University Bochum, Germany. In this section, the ODC preparation, measurements and the experimental setups are summarized shortly. A detailed description is given in the appendix C.

[8] Measurements performed by Institute of Chemical and Electrochemical Process Engineering, Technische Universität Clausthal, Germany and Analytical Chemistry — Center for Electrochemical Sciences, Ruhr University Bochum, Germany, published in [2].

In both setups polarization curves as well as potential steps were measured in a 10 M NaOH electrolyte at 50 °C. The main factor differentiating both setups was the mass transport at the liquid side offered in the respective system. While the electrolyte was driven by a convective flow in the first setup (con), it was stagnant in the second one (stag). In the first cell, the electrolyte temperature was controlled by regulating the temperature of the external electrolyte reservoir. In the second cell, an external heating circle had to be installed to enable temperature control without electrolyte flow. Furthermore, both setups differed in their geometrical cross section and in their electrode orientation respective to the ground (cf. fig. 5.1). A detailed documentation of the setup with the convective electrolyte flow is given in ref. [63], and for the cell with the stagnant electrolyte in ref. [34]. In the latter setup, SECM was used to quantify the hydroxide activity under stationary operating conditions.

Table 5.1: Electrode characteristics.

Name	Symbol	Unit	Convective electrolyte cell	Stagnant electrolyte cell
Electrode thickness	$z_{\text{electrode}}$	m	300×10^{-6}	295×10^{-6}
Geometrical cross section	A_{geo}	cm^2	3.14	0.264
Catalyst loading	θ_{Ag}	$\text{mg}\,\text{cm}^{-2}$	130.0	129.9
PTFE loading	θ_{PTFE}	$\text{mg}\,\text{cm}^{-2}$	4.0	4.1
OCP (measured)	E_0	V vs. RHE	1.095	1.10

5.3 Modelling and parameterization

In this section, first the expansion of the model is described. Subsequently the parameterization of the model using the dynamic and staionary measurements is given.

5.3.1 Expanded liquid mass transport model

The equations of the basic model have been introduced in the previous chapter 4. Since it is known from the last chapter, that ion concentration and water activity are crucial for the ODC performance, their influences are investigated by means of the expanded liquid mass transport model. Therefore following extensions have been made:

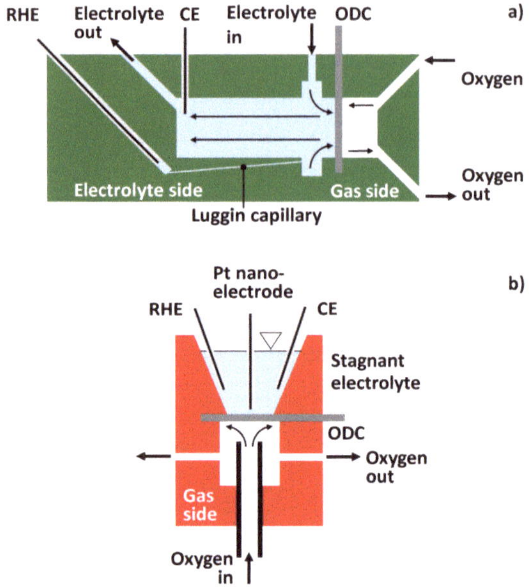

Figure 5.1: Schematic view of the used measuring cells with a) convective electrolyte flow and b) stagnant electrolyte.

- The ion concentration, water activity and also the oxygen availability affect the thermodynamics of the ODC and thus OCP. Considering this, the Nernst-equation (see eq. (5.4)) has been included into the model.
- To solve the Nernst-equation, the NaOH activity a_{NaOH} is needed. It is calculated based on the mean activity coefficient $\gamma_{\pm}$ of the electrolyte, assuming electro-neutrality, cf. eq. (5.6) and (5.7) [66].
- The oxygen activity a_{O_2} is modeled to be non-ideal and have been set into dependency of the ion concentration of the electrolyte, cf. eq. (5.5) [50].

The differences to the basic model are also summarized in table 3.1. All equations implemented in the expanded liquid mass transport model are given in table 5.2.

Table 5.2: Model equations for expanded liquid mass transport model. (Model schema for reference: fig: 4.1.)

Electrochemical reaction kinetics, double layer capacitance and internal resistance		
Faraday current	$j_F = -F \cdot 4 \cdot r$	(5.1)
Reaction rate	$r = k_0 \cdot a_{O_2} \cdot a^2_{H_2O} \cdot \exp\left(-\frac{(1-\alpha) \cdot F \cdot \eta_{reaction}}{R \cdot T}\right)$	(5.2)
Reaction overpotential	$\eta_{reaction} = E_{int} - E_0$	(5.3)
Nernst equation	$E_0 = E_{00} + \frac{R \cdot T}{4 \cdot F} \ln\left(\frac{a_{O_2} \cdot a^2_{H_2O}}{a^4_{NaOH}}\right)$	(5.4)
Activity of O_2	$a_{O_2} = x_{O_2} \cdot \gamma_{O_2}$	(5.5)
Activity of NaOH	$a_{NaOH} = \frac{b_{NaOH}}{b^{\Theta}_{NaOH}} \cdot \gamma_{\pm}$	(5.6)
Molality of NaOH	$b_{NaOH} = \frac{n_{NaOH}}{m_{NaOH}}$	(5.7)
Ohmic drop	$E_{ext} - E_{int} = R_{specific} \cdot j$	(5.8)
Charge balance	$C_{dl} \frac{dE_{int}}{dt} = j - F \cdot 4 \cdot r$	(5.9)
Mass transport in the gas phase and processes at the gas-liquid interface		
Species balance	$\frac{\partial p_i}{\partial t} \cdot \varepsilon = \frac{\partial^2 p_i}{\partial z^2} \cdot \mathscr{D}^{gas,eff}_{i,j} + \frac{\partial p_i}{\partial z} \cdot v_{gas}$ with $i \in \{H_2O, O_2\}$	(5.10)
Effective diffusion coefficient	$\mathscr{D}^{gas,eff}_{i,j} = \mathscr{D}^{gas}_{i,j} \cdot \frac{\varepsilon}{\tau}$	(5.11)
Bruggemann correlation	$\tau = \varepsilon^{-0.5}$	(5.12)
Henry's law	$c^*_{O_2} = \frac{p^*_{O_2}}{H_{O_2}(c^*_{NaOH}, c^*_{H_2O})}$	(5.13)
Stefan flow	$\frac{p_{total}}{R \cdot T} \cdot v_{gas} = \dot{N}^{evaporation}_{H_2O} - \dot{N}^{dissolution}_{O_2}$	(5.14)
Specific phase interface area	$S_{phase,interface} = \frac{A_{phase,interface}}{A_{geometric}}$	(5.15)
Mass transport in the thin film		
Species balance	$\frac{\partial c_i}{\partial t} = \sum_{\substack{j=1 \\ i \neq j}}^{n} \left(\frac{\partial^2 c_i}{\partial z^2} \frac{c_j}{c_{total}} - \frac{\partial^2 c_j}{\partial z^2} \frac{c_i}{c_{total}}\right) \cdot Đ^{liq}_{i,j} + \frac{\partial c_i}{\partial z} \cdot v_{liq,evaporation}$ with $i \in \{OH^-, H_2O, O_2\}$, $j \in \{OH^-, H_2O\}$	(5.16)

Local velocity due to evaporation	$\frac{\mathrm{d}v_{\text{liq,evaporation}}}{\mathrm{d}z} = 0$ (5.17)
Mass transport in the flooded agglomerates	
Species balance	$\frac{\partial c_i}{\partial t} \cdot \varepsilon = \sum_{\substack{j=1 \\ i \neq j}}^{n} \left(\frac{\partial^2 c_i}{\partial z^2} \frac{c_j}{c_{\text{total}}} - \frac{\partial^2 c_j}{\partial z^2} \frac{c_i}{c_{\text{total}}} \right) \cdot Đ_{i,j}^{\text{liq,eff}} + \frac{\partial c_i}{\partial z} \cdot v_{\text{liq}} - r$
	with $i \in \{OH^-, H_2O, O_2\}$, $j \in \{OH^-, H_2O\}$ (5.18)
Velocity due to reaction	$v_{\text{liq,react}}(z) = (\lvert \nu_{H_2O}^{\text{ORR}} \rvert - \lvert \nu_{OH^-}^{\text{ORR}} \rvert) \int_0^z \frac{r(z)}{c_{\text{total}}} \mathrm{d}z$ (5.19)
Total velocity of liquid phase	$v_{\text{liq}}(z) = v_{\text{liq,evaporation}} + v_{\text{liq, react}}(z)$ (5.20)
Mass transport in the liquid diffusion layer	
Species balance	$\frac{\partial c_i}{\partial t} = \sum_{\substack{j=1 \\ i \neq j}}^{n} \left(\frac{\partial^2 c_i}{\partial z^2} \frac{c_j}{c_{\text{total}}} - \frac{\partial^2 c_j}{\partial z^2} \frac{c_i}{c_{\text{total}}} \right) \cdot Đ_{i,j}^{\text{liq}} + \frac{\partial c_i}{\partial z} \cdot v_{\text{ldl}}$ (5.21)
	with $i \in \{OH^-, H_2O, O_2\}$, $j \in \{OH^-, H_2O\}$
Velocity	$v_{\text{ldl}} = v_{\text{liq}}(z = z_{\text{l}})$ (5.22)
Boundary conditions	
O_2 in gas bulk	$p_{O_2}(z = 0) = p_{\text{total}}$ (5.23)
H_2O in gas bulk	$p_{H_2O}(z = 0) = 0$ (5.24)
O_2 in electrolyte bulk	$c_{O_2}(z = z_{\text{t}}) = 0$ (5.25)
NaOH in electrolyte bulk	$c_{\text{NaOH}}(z = z_{\text{t}}) = c_{\text{NaOH}}(z = z_{\text{t}}, t = 0)$ (5.26)
H_2O in electrolyte bulk[a]	$c_{H_2O}(z = z_{\text{t}}) = f(c_{\text{NaOH}}(z = z_{\text{t}}, t = 0))$ (5.27)
Electrode geometry	$z_{\text{gas}} + z_{\text{tf}} + z_{\text{fa}} = z_{\text{electrode}}$ (5.28)

[a] Calculated as in [51]

5.3.2 Parameterization and parameter evaluation

Constant model parameters are adjusted to the operating conditions of the measurements and are listed in table 5.3, concentration dependent parameters are given in table 5.4. For the manual parameter identification, dynamic potential steps from 0.8 to 0.7 V vs. RHE and back to 0.8 V as well as an experimental polarization curve coupled with concentration measurements were used. The experimental data and simulation results will be discussed in the next section. The agreement between experiment and simulation is satisfactory for the polarization curves as well as the potential steps. The parameter values for both cells are presented in table 5.5. The combination of stationary and dynamic measurements enables one to cover a wide operation range but also to determine the different processes in the electrode separately due to their characteristic time constants.

As expected, significantly different values for both systems are obtained for the length of the liquid boundary layer z_{ldl} in and the electrolyte bulk phase: $z_{ldl}^{con} = 10\,\mu m$ for the setup with the convective electrolyte flow and $z_{ldl}^{stag} = 270\,\mu m$ for the setup with the stagnant electrolyte. For the reaction rate constant k_0, the specific interface $S_{phase,interface}$ and the average length of flooded agglomerates z_{fa}, both systems exhibits values in the same order of magnitude. The larger specific interface in the cell with the convective electrolyte flow may occur due to the higher motion in the liquid electrolyte and is associated with temporary local shifts of the gas-liquid interface on the pore scale level.
The large differences in double layer capacitances may be explained by its dependency on the current densities [72], which differs considerably in both systems at the respective potentials. Additional possible measurement inaccuracies are discussed below.
In my previous work dealing with an ODC operated at 80 °C, I presented a higher value of $S_{phase,interface} = 134\,m^2\,m^{-2}$ and a slightly higher value for z_{fa} of $11.5\,\mu m$ [1]. On the one hand this deviation may originate from temperature dependent system characteristics like viscosity or wettability. On the other hand, the use of dynamic measurements in my present work allows for a better separation and a more precise identification of the process parameters.

5.4 Steady state analysis of oxygen depolarized cathodes performance

Figure 5.2 displays the measured and the simulated polarization curves of both systems. In the kinetically governed regime at small current densities both cells have almost the same behavior, which is expected, since similar electrodes were used. Accordingly, the derived reaction rate constants k_0 (cf. table 5.5) are in the same order of magnitude.
In contrast, at medium current densities up to the diffusion limited current density, a strong difference between both systems occurs: While the system with the stagnant electrolyte shows a significant mass transport limitation at about $j \approx 1.3\,kA\,m^{-2}$ at $E_{Ext} = 0.1\,V$ vs. RHE, no mass transport limitation proceed for the system with electrolyte convection at $j = 3.8\,kA\,m^{-2}$ at the

Table 5.3: List of constant parameters and operating conditions.

Name	Symbol	Unit	Value
Electrode geometry			
Porosity [32]	ε	–	0.4
Operating conditions			
Pressure in gas chamber[a]	$p = p_{O_2}$	Pa	1.013×10^5
NaOH concentration in electrolyte bulk[a]	$c_{NaOH}(z = z_t)$	$mol\,m^{-3}$	1.00×10^4
H_2O concentration in electrolyte bulk [51]	$c_{H_2O}(z = z_t)$	$mol\,m^{-3}$	5.05×10^4
Temperature[a]	T	K	323.15
Electrochemical and kinetic data			
Charge transfer coefficient[b] [24]	α	–	0.15
Stoichiometric coefficient H_2O	$\nu_{H_2O}^{ORR}$	–	−2
Stoichiometric coefficient OH^-	$\nu_{OH^-}^{ORR}$	–	+4
Binary diffusion coefficients			
For gaseous oxygen/water [73]	$\mathscr{D}_{O_2,H_2O}^{gas}$	$m^2\,s^{-1}$	2.48×10^{-5}
For liquid oxygen/water[c] [73, 74]	$Đ_{O_2,H_2O}^{liq}$	$m^2\,s^{-1}$	1.36×10^{-9}
For liquid oxygen/NaOH[c] [73, 74]	$Đ_{O_2,NaOH}^{liq}$	$m^2\,s^{-1}$	2.02×10^{-9}
For liquid water/NaOH[d] [60]	$Đ_{H_2O,NaOH}^{liq}$	$m^2\,s^{-1}$	5.33×10^{-10}

[a] Experimental operating condition. [b] Measured with 6.5 M NaOH. [c] Calculated with modified Wilke and Chang equation, with $\Theta = 3.9$ and $a = 0.5$ chosen, adjusted to experimental data from [74]. [d] No NaOH data available, modeled for KOH with NaOH viscosity from [68], as seen in [32].

Table 5.4: Electrolyte concentration dependent variables.

Name	Symbol	Unit
Inverse Henry constant[a] [22]	H_{O_2}	$Pa\,m^3\,mol^{-1}$
Activity coefficient of O_2 [b] [50]	γ_{O_2}	–
Activity of H_2O [75]	a_{H_2O}	–
Mean activity coefficient of NaOH [66]	$\gamma_{\pm}$	$kg\,mol^{-1}$
Density of electrolyte [68]	ρ	$kg\,m^{-3}$
Water vapor pressure [51]	$p_{H_2O}^{vap}$	Pa

[a] Extrapolated from $c_{NaOH} \leq 6\,M$, but as shown in [27] extrapolated values fit very well to experiments with $c_{NaOH} \leq 12\,M$, [b] Extrapolated from $c_{NaOH} \leq 5.5\,M$, but as shown in original research, calculations for e.g. $c_{KOH} \leq 13.5\,M$ are valid [50].

same potential.

Table 5.5: Parameters identified from electrochemical experiments.

Name	Symbol	Unit	Convective electrolyte	Stagnant electrolyte
Reaction rate constant	k_0	$mol\,s^{-1}\,m^{-2}$	15.42×10^4	9.20×10^4
Specific interface	$S_{phase,interface}$	$m^2\,m^{-2}$	56.0	36.4
Liquid diffusion layer thickness	z_{ldl}	m	10×10^{-6}	270×10^{-6}
Overall specific resistance	$R_{specific}$	$\Omega\,m^2$	1.16×10^{-4}	1.82×10^{-4}
Length of flooded agglomerates	z_{fa}	m	7.90×10^{-6}	8.14×10^{-6}
Thin-film thickness [a]	z_{tf}	m	80×10^{-9}	
Double layer capacitance	C_{dl}	$F\,m^{-2}$	80	720

[a] Set as equal for both systems. Typical values for the thin-film thicknesses: 25 - 300 nm [45].

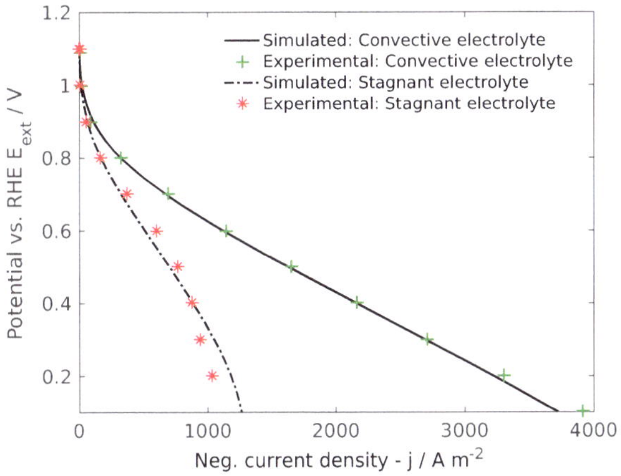

Figure 5.2: Simulated polarization curves in comparison to experimental polarization curves for convective electrolyte flow and stagnant electrolyte cell.

The ORR can proceed at locations within the flooded agglomerates where the reactants oxygen and water both are present. In the following, this section is called reaction zone. In figure 5.3, simulated oxygen concentration profiles over the thin-film and the first micrometer of the flooded agglomerates for both systems at different current densities are shown. For both systems the reaction zone, facing the gas-liquid interface, decreases in size with increasing current densities due to the depletion of oxygen. Similarly to what was presented earlier [1], no oxygen associated limitation takes place due to the mass transport in the gas phase, therefore all oxygen mass transport limitation can be attributed to the liquid phase.

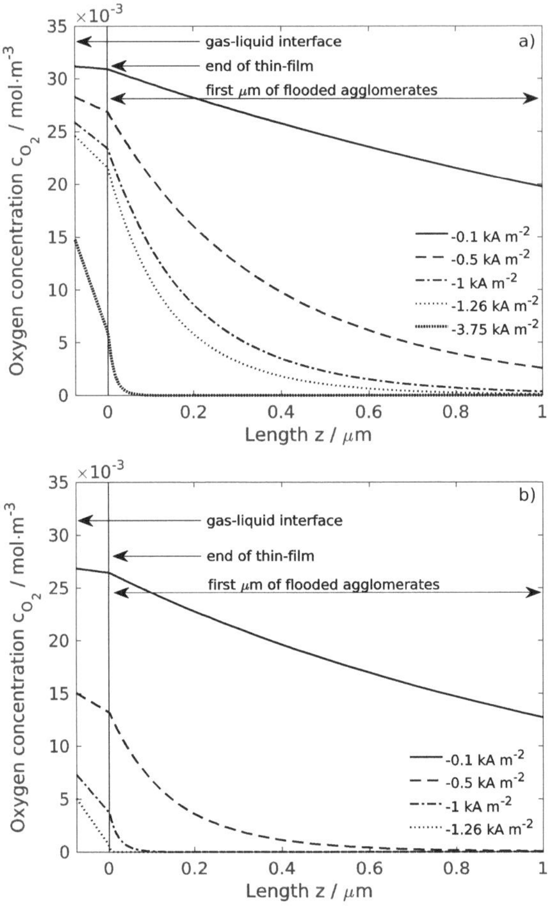

Figure 5.3: Dependence of simulated oxygen concentrations profiles within the liquid phase with respect to the current density for a) convective electrolyte flow and b) stagnant electrolyte.

In the system with the stagnant electrolyte, the reaction zone shrinks to almost zero at the mass transport governed current density of $j \approx 1.3\,\mathrm{kA\,m^{-2}}$. This shows that the limiting current density is determined by the availability of the oxygen. In contrast, for the system with the convective electrolyte flow, the reaction zone amounts to several nanometers even at a current density of $j = 3.8\,\mathrm{kA\,m^{-2}}$ so that no oxygen mass transport limitation occurs.

The vast disparity in oxygen availability is caused by the difference of current dependent water activity inside the electrodes. The ORR consumes water and produces hydroxide ions within the reaction zone close to the gas-liquid interface. Since three to four water molecules form a solvation shell around the hydroxide ion [78], the amount of free water molecules and thus the water activity decreases significantly due to the production of hydroxide ions. This change of water

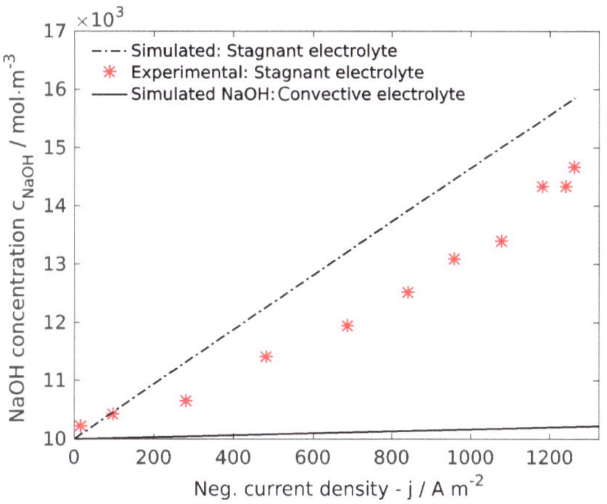

Figure 5.4: Simulated and experimental NaOH concentration in the liquid diffusion layer 1 µm above the ODC surface for cell with stagnant electrolyte. For comparison, NaOH concentration in the cell with the convective electrolyte flow is also given at the same location.

activity leads to a change of the phase equilibrium of water and oxygen, and thus to a reduced oxygen solubility [50] and a higher vapor pressure of water [51]. There is also experimental evidence for the accumulation of hydroxide ions in the system with the stagnant electrolyte: Figure 5.4 shows the hydroxide ion concentration detected via SECM 1 µm above the ODC surface (cf. fig. 5.1), in comparison to the simulated values at the same location. It can be deduced that both, the measurement as well as the model, show a significant increase of hydroxide concentration and subsequently a decrease of the water activity. Differences between experiment and simulation amounts to maximum 10 % and might result from additional convection caused by a density gradient – due to temperature or ion concentration gradients – or local inhomogeneities of ion concentration in-plane direction of the electrode in the experiment [34]. The mass transport resistance for the exchange of water and hydroxide ions between the ODC and the electrolyte bulk is much higher for the thick liquid diffusion layer in the system with the stagnant electrolyte than in the system with the convective electrolyte flow. It may be that the mass transport in the liquid diffusion layer is not only influenced by the forced convection, but also by the vertical orientation of the ODC. A thicker diffusion layer constitutes a larger diffusion resistance and causes a higher accumulation of hydroxide ions in the electrode. Thus, the accumulation of the hydroxide has a detrimental effect on the oxygen solubility, which is higher in the system with the stagnant electrolyte than in the system with the convective electrolyte flow. This in turn explains why the system with the stagnant electrolyte is limited by oxygen availability at $E = 0.1$ V vs. RHE.

5.5 Dynamic analysis

In this section the influence of the reaction-induced oxygen solubility limitation on the dynamics of the ODCs is discussed in detail based on dynamic measurements and simulations of both configurations. In figure 5.5, the current response to a potential step from 0.8 to 0.7 V vs. RHE and back is displayed for both systems.

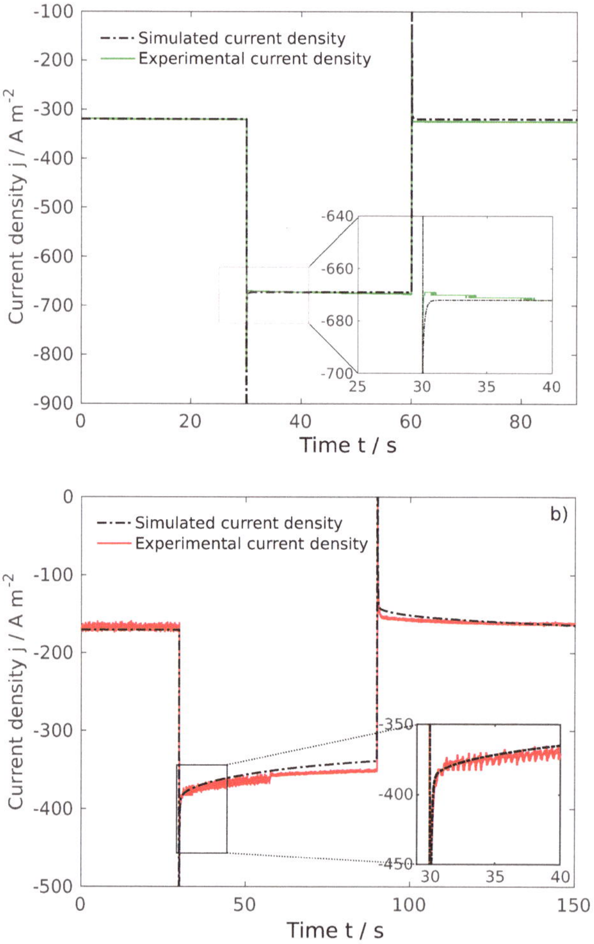

Figure 5.5: Simulated and experimental time-dependent current density response for a potential step from 0.8 to 0.7 V vs. RHE at $t = 30$ s and a) back at $t = 60$ s for convective electrolyte flow cell and b) back at $t = 90$ s for stagnant electrolyte cell.

Although the same potential is applied to both electrodes, the system with electrolyte convection offers a higher cathodic current density, which agrees with the polarization curves and has been discussed in the previous section. Furthermore, both systems exhibit a very different dynamic response to the potential step, which will be discussed individually. Two distinct processes can be separated: The charging and discharging of the double layer and the mass transport of water and hydroxide ions across the liquid phase. As discussed below, the dynamics of the oxygen follow the latter process.
The charging and discharging of the double layer is the reason for the positive and negative current peaks directly after the positive and the negative potential step. For both systems, different time constants are obtained, which is attributed to the dependency of the double layer capacitance on the current density [72]. Additional a lager area outside the reactive area due to small leakages between the sealing ring and the ODC of the double layer may be penetrated in the smaller cell with the stagnant electrolyte. Since the electrode circumference / area ratio is approx. 3.5 times larger for the smaller cell than for the lager cell with the convective electrolyte flow, this effect may be more significant.
The subsequent relaxation of the current density occurring in the range of seconds is dominated by the mass transport of water and hydroxide ions between the reaction zone and the electrolyte bulk phase [1]. The corresponding time constant can be approximated with equation (3.27), derived from Fick's second law:

$$\tau_{\mathrm{mt}} \approx \frac{\Delta z^2}{D} \tag{5.29}$$

With τ_{mt} as the time constant of water and hydroxide ion mass transport, D as the diffusion coefficient and Δz as the mass transport length. For both systems the mass transport time constant is a function of the diffusion through the porous structure of the flooded agglomerates and the free diffusion through the liquid diffusion layer (cf. fig. 4.1), which can be calculated separately. Since the length of the flooded part of both electrodes is almost the same, similar time constants are obtained: $\tau_{\mathrm{mt,H_2O}}^{\mathrm{fa,\,con}} = 0.19\,\mathrm{s}$ for the convective electrolyte flow and $\tau_{\mathrm{mt,H_2O}}^{\mathrm{fa,\,stag}} = 0.20\,\mathrm{s}$ for the stagnant electrolyte. In contrast, the time constants for the mass transport across the liquid diffusion layer differs by orders of magnitude between both systems: $\tau_{\mathrm{mt,H_2O}}^{\mathrm{ldl,\,con}} = 0.19\,\mathrm{s}$ and $\tau_{\mathrm{mt,H_2O}}^{\mathrm{ldl,\,stag}} = 181.3\,\mathrm{s}$. The large time constant is in line with the observation that no steady state is reached within the 60 s duration of the dynamic test for the system with the stagnant electrolyte, but within 1 s for the system with the convective electrolyte flow. In conclusion, the water and hydroxide ion transport across the liquid diffusion layer dominates the current response to the potential step and causes the large difference in the current response of both electrodes. The corresponding time-dependent ion concentration and water activity at the gas-liquid interface and 1 µm away from the ODC surface are presented in figure 5.6.

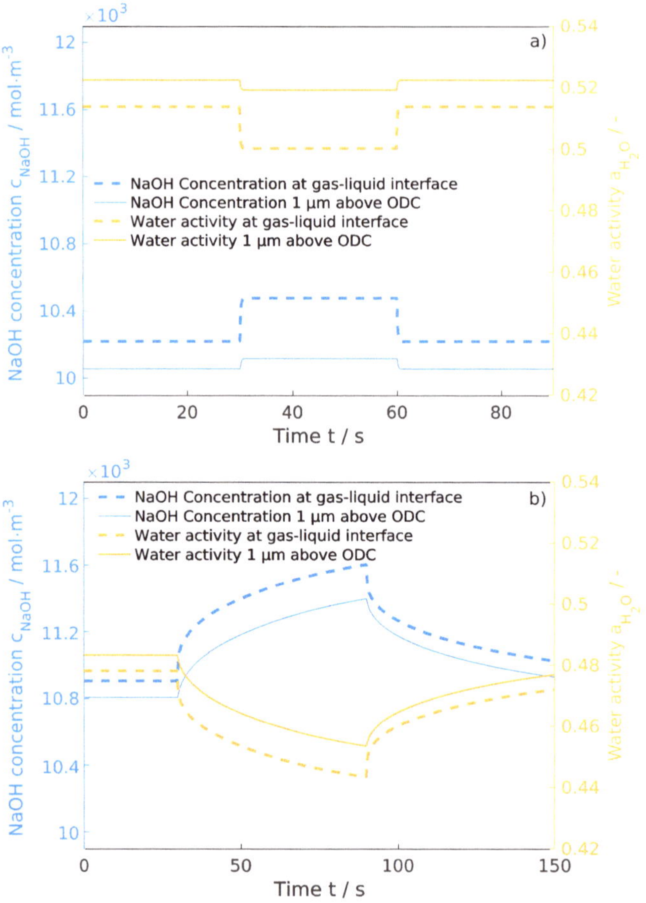

Figure 5.6: Simulated NaOH concentration and water activity within the liquid phase dependent on the time for a potential step from 0.8 to 0.7 V vs. RHE at $t = 30$ s and a) back at $t = 60$ s for convective electrolyte flow cell and b) back at $t = 90$ s for stagnant electrolyte cell.

Due to the relationship between water and hydroxide activity as discussed above, the activities of both species show opposite trends. Furthermore, water activity has a strong influence on oxygen solubility [75]. This effect is illustrated in figure 5.7 for the system with the stagnant electrolyte. The right-hand side of the plot displays the steady state water activity gradient within the liquid diffusion layer for $t = 30$ s directly before the negative potential step and at selected time points afterwards. The plot in the middle shows the same process within the flooded agglomerates. A comparison of the curves suggests that the gradient over the whole electrode and the liquid diffusion layer change with a delay. This is due to the fast reaction rate compared to slow mass transport of water and hydroxide ions. The plot on the left-hand side of figure 5.7 depicts the

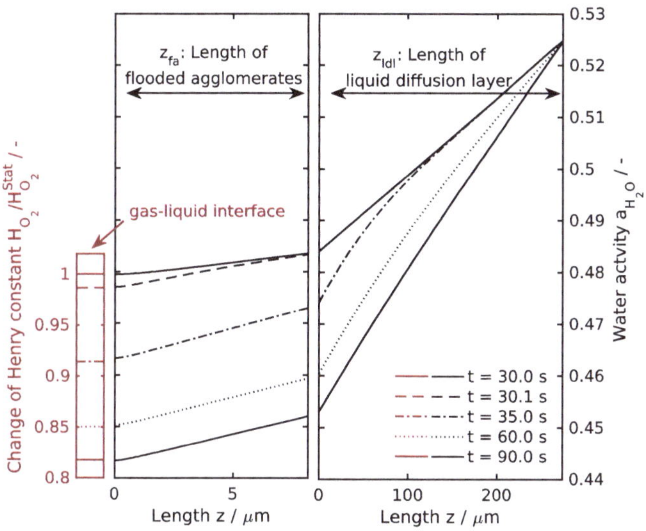

Figure 5.7: Henry constant at the gas-liquid interface (left) at steady state current at 0.7 V vs. RHE and its change at selected time points after a potential step to 0.8 V vs. RHE. Corresponding gradient of water activity within the ODC across the flooded agglomerates (middle) and in the liquid diffusion layer (right). Cell with no electrolyte convection.

corresponding change of oxygen solubility for the different points in time, indicated by the relative change of the Henry constant at the gas-liquid interface during the step. The change of the Henry constant directly affects the oxygen solubility. In the cell with electrolyte convection the liquid diffusion layer is smaller and the new steady state of the phase equilibrium and thus of oxygen concentration is reached much faster.

The resulting response of oxygen concentration with and without electrolyte convection is depicted in figure 5.8 for different locations. It can be inferred that due to the ion dependent solubility the oxygen concentration is governed by the same time constant as the ion concentration. Therefore, the oxygen supplied to the reaction zone and thus the current density exhibits the same time constants as the liquid phase mass transfer.

In figure 5.5, deviations between experiments and simulations are visible. For the system with electrolyte convection, a slight decrease of the negative current density with time after the negative potential step at 30 s can be identified in the experiment. For the system without electrolyte convection, the measured current density is more cathodic than the simulated one. Different factors which are not included in the model might have contributed to this behavior: Although the experiments were conducted under temperature control, a moderate increase of the ODC temperature which would improve kinetics and transport processes, cannot be excluded. In that regard, heating effects can be caused by ohmic losses or the entropy effect, for instance [8]. The same effects may occur for the steady state results shown above.

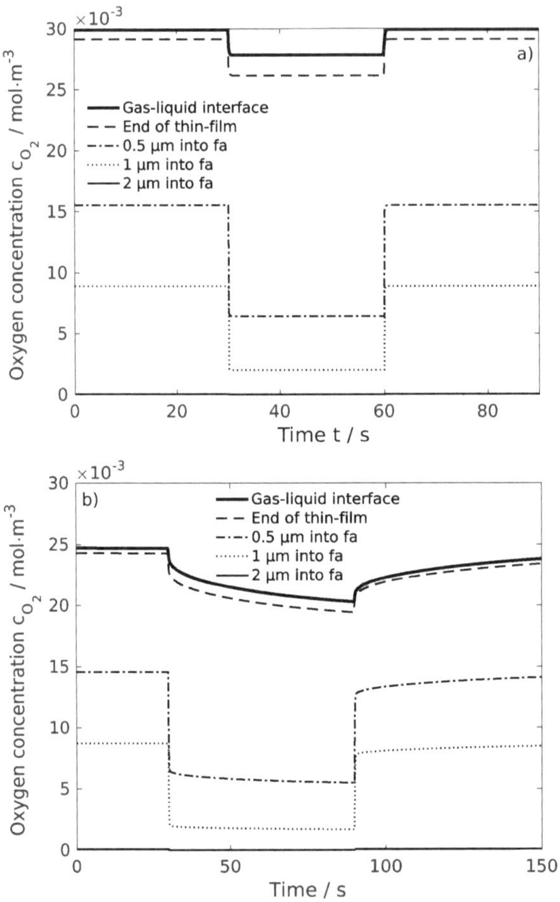

Figure 5.8: Simulated oxygen concentration within the liquid phase dependent on the time for a potential step from 0.8 to 0.7 V vs. RHE at $t = 30$ s and a) back at $t = 60$ s for convective electrolyte flow cell and b) back at $t = 90$ s for stagnant electrolyte cell (fa: flooded agglomerates).

From the dynamic analysis, I can conclude that the performance as well as the dynamics of ODCs are determined by the mass transport across the liquid phase. This insight is corroborated by two main aspects: On the one hand, in the system with the stagnant electrolyte, water activity decreases only by less than 7.5 % as a consequence of the negative potential step, but oxygen solubility decreases by 18 % (cf. fig. 5.7). This suggests a minor influence of the water activity on the kinetics in a direct way but a high indirect impact due to the oxygen solubility. On the other hand, due to the faster water mass transport, the oxygen solubility of the system with the convective electrolyte flow only decreases by 7 %. The comparison of both systems indicates that the mass transport within the liquid electrolyte has a significant influence on the electrode performance. It is expected that also electrode design and temperature will have significant impact, as they are known to affect transport phenomena. This leads to the conclusion that the performance of technical ODCs can be improved by accelerating the mass transfer in the liquid phase, e.g. via convective electrolyte flow or electrolyte flow field optimization. Already now, in industrial application quite high performances are reached due to convective electrolyte transport in so-called percolators [3], further performance increase might be reached by a even further optimized mass transport.

5.6 Conclusions

Experimental polarization curves and potential steps demonstrate that the performance and dynamic behavior of ODCs deviate strongly when the electrodes are operated with and without convection of the liquid electrolyte. Without electrolyte convection, a mass transfer limitation is reached at a current density of $j \approx 1.3\,\mathrm{kAm^{-2}}$, whereas this does not occur at current densities of $j > 3\,\mathrm{kAm^{-2}}$ with electrolyte convection. Furthermore, the dynamic response of the current density to a potential step exhibits a significantly slower relaxation behavior without electrolyte convection.

By comparing both systems utilizing simulation and measurements, the hypothesis that water activity plays a key role was confirmed. The accumulation of hydroxide ions leads to a reduction in water activity and thus in oxygen solubility. Since three to four water molecules are bound to every hydroxide ion as a solvation shell, the water activity decreases rapidly as a function of the ionic strength, and the accompanying change of the phase equilibrium leads to a lower oxygen solubility. The model-based analysis suggests that electrolyte convection increases mass transfer rates in the liquid phase which in turn reduces the hydroxide ion accumulation and thereby increases both, water activity and oxygen solubility. Without electrolyte convection, a thicker liquid diffusion layer is established causing a stronger accumulation of hydroxide ions and a lower water activity. Not all processes can be explained by mass transport and phase equilibrium phenomena, so that further research on catalyst activity and local reaction rates within porous structures is needed.

In conclusion, a high mass transport rate of water and hydroxide ions is essential for high ODC performance and needs to be ensured for technical applications of these kind of electrodes. A thorough understanding of the transport processes is also indispensable for understanding the dynamic behavior of the electrodes.

6 In-depth Analysis of Oxygen Depolarized Cathodes Structure and Electrolyte Distribution[9]

In this chapter, a modeling approach of ODCs to evaluate the influence of inhomogeneities in electrolyte distribution on electrode performance and dynamics is presented. It is investigated how the location and size of the gas-liquid interface affect the local reaction rates within the ORR.

6.1 Introduction

Recently, Franzen et al. [23] analyzed in their experimental study ten ODCs with different compositions and structures, and found significant performance differences. In detail, they evaluated ODCs with a silver content of 90 to 99 wt−% (10 to 1 wt−% PTFE, respectively) and identified a silver content of 98 wt−% as best. The amount of silver had a very low influence on the polarization curve at low, e.g. kinetically governed current densities, but a significant influence on the performance at high current densities, i.e. in the mass transport dominated range of the polarization curve. It was hypothesized that the hydrophilic ODC with a high silver content of 99 wt−% was flooded by the liquid electrolyte, whereas ODCs with a high PTFE content above 5 wt−% were too hydrophobic and electrolyte intrusion was insufficient. Both effects may lead to a decrease in the extent of the gas-liquid interface and thus an insufficient oxygen availability in the liquid phase [23].

Paulisch et al. [80] investigated the electrolyte distribution within ODCs by operando X-ray radiography and chronoamperometric measurements. They showed that the pore structure as well as the distribution of the electrolyte during operations is inhomogeneous: some pores were barely flooded by the electrolyte, whereas a local breakthrough of the electrolyte to the oxygen side was observed in other pores [80]. These findings suggest that the gas-liquid interface may also be inhomogeneously distributed over the entire thickness of the electrode. The identified locally

[9] Parts of this chapter have been published in Röhe et al., Electrochimica acta, 2021 [15].

high electrolyte saturation seems surprising, as previous studies have shown that even very small amounts of PTFE lead to a non-wettability of a pore system [81].
So far in this dissertation, it was demonstrated that the ODC performance is limited by the availability of dissolved oxygen in the liquid phase, whereby the oxygen solubility rate is significantly affected by the length of the water and hydroxide mass transport path between the gas-liquid interface and the liquid bulk.
The findings of Franzen et al., Paulisch et al. and the results of this work raise new questions: How does the phase distribution affect the ODC dynamics and performance? Does the distributed mass transport length of water and ions in the liquid phase lead to locally different oxygen solubilities and thus to different local reaction rates? Is this effect related to the silver / PTFE ratio dependent ODC performance? To answer these open-ended questions and to quantify the influence of the electrolyte distribution, a model is needed that considers the inhomogeneities in electrolyte distribution within ODCs. The importance of understanding the distribution of the liquid phase in porous electrodes was already emphasized by various experimental and stimulative studies inter alia for proton-exchange membrane fuel cell (PEMFC) [82–84].
Model-based studies to investigate inhomogeneities were inter alia performed by Levi et al. who connected two different diffusion paths in parallel to study diffusion time constants for particle size distributions in non-uniform composite graphite electrodes for lithium-ion batteries [85] or by Jüttner and Lorenz who investigated the influence of surface inhomogeneities on corrosion processes for different metal electrodes in sodium hydroxide electrolyte [86]. Chevalier et al. interconnected single cell models to draw conclusions about the local distribution of the state-of-health of a PEMFC stack from the stack impedance spectra [36]. Krewer et al. have shown that local inhomogeneities in concentration in direct methanol fuel cells (DMFC) lead to characteristic changes of the dynamic behavior [87].
In this chapter, I present the first model for describing the influence of inhomogeneities in electrolyte distribution on ODC performance. To this end, the in chapter 5 introduced dynamic expanded liquid mass transport model is extended to incorporate multiple electrode domains, each with different representative characteristics. The inhomogeneous ODC model is used to analyze experimental electrochemical impedance spectra (EIS) as well as polarization curves[10] of ODCs with systematically varied silver catalyst content. It will be shown how to interpret the observed flattened semicircles in the impedance spectra regarding to local inhomogeneities. I demonstrate how the structure and silver content of ODCs affect the electrolyte distribution, electrode performance and dynamics. The aim of this study is to infer the macroscopic distribution of the electrolyte from the performance and impedance spectra of the ODCs. The methodology

[10] Measurements performed by Institute of Chemical and Electrochemical Process Engineering, Technische Universität Clausthal, Germany, published in [15].

used does not allow conclusions to be drawn about microstructures such as pore size distribution or local electrolyte distributions. It will be discussed that not only the size of the gas-liquid interface but also the location is of importance for the performance. These are key findings to understand the influence of the electrode structure on the electrode performance and dynamics. EIS is shown to be a highly sensitive tool to diagnose the electrolyte distribution within ODCs, while this information is hardly accessible with potential steps or polarization curves.

6.2 Experimental

The measurements have been performed by Institute of Chemical and Electrochemical Process Engineering, Technische Universität Clausthal, Germany. Detailed descriptions of the experiments and also the electrode preparation are given in the appendix D and by Franzen et. al [23]. In short: The electrochemical experiments have been performed in a Gaskatel half-cell (FlexCell HZ PP01, Gaskatel GmbH) with a geometrical area of $3.14\,\mathrm{cm}^2$. After a start-up procedure, pseudo-galvanostatic impedance measurements have been performed at current densities of $2\,\mathrm{kA\,m}^{-2}$ with a $10\,\mathrm{mV}$ amplitude. Subsequently polarization curves have been measured by linear sweep voltammetry, starting at open cell potential (OCP) with a scan rate of $0.5\,\mathrm{mV\,s}^{-1}$. Chemical characterizations of the ODCs are given in [23].

6.3 Analysis of measurement results

In this section, the measurement results which will be used for validating the simulations are explained and interpreted. As discussed in detail by Franzen et al., the Ag ratio of ODCs has a major effect on the stationary performance of ODCs [23]. Thus, the polarization curves of four ODCs with distinct silver-PTFE ratio have been chosen for further analysis from previously published data [23]:

- $92\,\mathrm{wt}-\%\,\mathrm{Ag}$: This composition exhibits a medium performance and hydrophobic properties (Electrode thickness: $292\,\mu\mathrm{m}$) [23].
- $97\,\mathrm{wt}-\%\,\mathrm{Ag}$: The structure of an ODC with $97\,\mathrm{wt}-\%\,\mathrm{Ag}$ has been analysed by Neumann et al. [69], so that identified structure parameter can be used in this study (Electrode thickness: $297\,\mu\mathrm{m}$ [23]).
- $98\,\mathrm{wt}-\%\,\mathrm{Ag}$: This composition exhibits the best performance [23] (Electrode thickness: $335\,\mu\mathrm{m}$).

- 99 wt−% Ag: This composition exhibits a low performance and hydrophilic properties [23] (Electrode thickness: 356 μm).

In the following, the ODCs are denoted by their silver content; the PTFE content is equal to rest of the mass fraction. The polarization curves of these ODCs are given in fig. 6.1. The main differences between the electrodes occur at current densities above 2000 A m^{-2}.

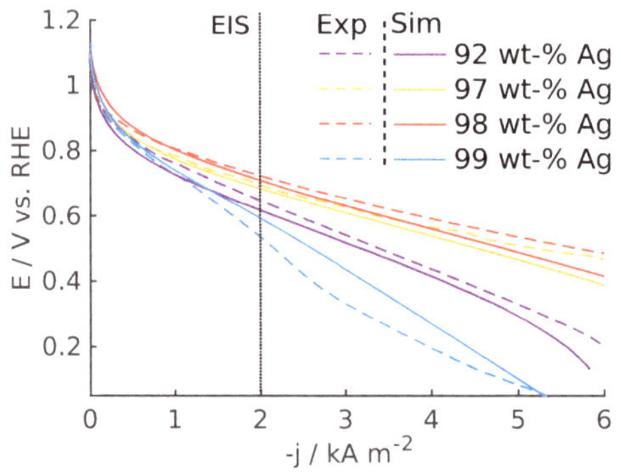

Figure 6.1: Experimental [23] and simulated polarization curves of ODCs with different Ag-content.

To explore the reasons of the different performances, electrochemical impedance measurements are conducted in this study, the results of which are given in fig. 6.2 a). All impedance spectra show distorted semi-circles. For the ODC with 99 wt−%, a second semicircle emerges at high frequencies. The ODCs mainly differ in terms of ohmic and charge transfer resistance. The ohmic and the charge transfer resistance R_{ct} of the different impedance spectra is determined as [62]:

$$R_{ct} = Z_R(f_{min}) - Z_R(f_{max}) \tag{6.1}$$

With Z_R as the real part of the impedance $|Z|$ and f as the frequency, where $Z_R(f_{max})$ is the ohmic resistance. There is no systematic correlation between ohmic resistance and silver content or performance of ODCs; the differences may result from slightly different distances between the ODCs and the reference electrode, since sealing elements are tightened by hand and therefore their compressions differ. The resulting values for R_{ct} are shown in fig. 6.2 b). The smallest charge transfer resistance is obtained for the ODC with 98 wt−%, the largest for 99 wt−%, which

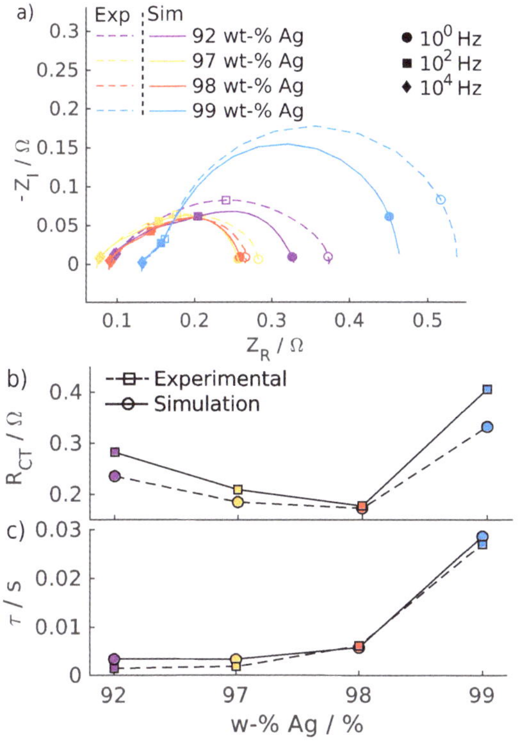

Figure 6.2: a) Simulated and experimental EIS at $j = 2000\,\mathrm{A\,m^{-2}}$ of ODCs with different Ag-content, b) calculated charge transfer resistance and c) calculated time constants from experimental and simulated EIS.

agrees with the performance ranking in stationary measurements. To identify the associated processes, the time constants were determined from the impedance data by eq. (6.2):

$$\tau = \frac{1}{2\pi f(-Z_{\mathrm{I,max}})} \tag{6.2}$$

$f(-Z_{\mathrm{I,max}})$ is the frequency where the negative imaginary part of the impedance reaches its maximum. The results are given in fig. 6.2 c). Two trends are visible: First, the time constant is increasing with an increasing Ag-content, but does not correlate with ODC performance. Second, for all four ODCs the characteristic time constant is in the range of 0.0015 s to 0.03 s, which agrees with previous results from potential step measurements [2]. It should be noted that the potential step measurements cover a larger range of time constants than EIS [35], and that no time constants slower than those found here were observed [2]. The charge transfer resistance reaches a minimum at 98 wt−% Ag, but the time constant decreases further until 92 wt−% Ag. In principle, the measured semi-circle and the corresponding time constants can be caused by transport processes, by the interplay of reaction and double layer charging, or a combination of both. However, mass transport can be excluded in this case. The time constant of a diffusion process τ_{mt} for species i in j can be approximated from Fick's second law [13]. Equation (6.3) gives the diffusion time constant within a porous media, which is proportional to the square of the characteristic diffusion length, i.e. the electrode thickness Δz, the porosity ε and inversely proportional to the binary diffusion coefficient $D_{i,j}^{\mathrm{eff}}$ of species i in j.

$$\tau_{\mathrm{mt}} \approx \frac{\Delta z^2}{D_{i,j}^{\mathrm{eff}}} \cdot \varepsilon \tag{6.3}$$

The diffusion processes in ODCs were analyzed in a previous chapter. As demonstrated, a long mass transport way in the liquid phase leads to a distinct time constant in chronoamperometric measurements due to the water and hydroxide ion mass transport, whereas no distinct time constants for oxygen transport – in the gas as well as in the liquid phase – were found [2]. For j as water and i as NaOH, a flooding degree of the ODC from 5 % to 50 % with $D_{i,j}^{\mathrm{eff}} = 3.20 \times 10^{-10}\ \mathrm{m^2\,s^{-1}}$ (cf. table 6.5) and $\varepsilon = 0.4$ would lead to time constants of 0.28 to 28.2 s, which is orders of magnitudes higher than the measurement results. Also, the results in fig. 6.2 suggest that the observed time constant is not determined by oxygen transport in the gas phase, since the time constants become faster with increasing hydrophilicity and thus decreasing mass transport length in the gas phase. Although the current density of ODCs is limited by the availability of dissolved oxygen in the liquid electrolyte [1, 2, 32], oxygen transport in the liquid phase can also be excluded as the process causing the measured semi-circles for two reasons: First, as shown before, the diffusion length of the oxygen in the liquid phase is only several tens nanometers [32, 88] up to a few hundred nanometers [89], and the associated time constant is smaller than 10×10^{-6} s [1]. This is orders of magnitudes faster than the time constants measured here. Second, the time constants of all ODCs differ while oxygen diffusivity in the electrolyte is not expected to be changed with the electrolyte distribution or electrode composition.
In shutdown measurements with comparable ODCs, double layer capacitances of about $C_{\mathrm{dl}} = 50\,\mathrm{F\,m^{-2}}$ were identified [72]. The time constant of the double layer charge / discharge taking place in parallel to a faradaic reaction is given by eq. (6.4) [62]:

$$\tau_{\mathrm{dl}} = C_{\mathrm{dl}} \cdot A_{\mathrm{geo}} \cdot R_{\mathrm{ct}} \tag{6.4}$$

With the characteristic time constant τ_{dl}, the double layer capacitance per geometrical area C_{dl}, and the geometrical electrode area A_{geo}. Using $C_{\mathrm{dl}} = 50\,\mathrm{F\,m^{-2}}$, $R_{\mathrm{ct}} = 0.2\,\Omega$ and $A_{\mathrm{geo}} = 6.28 \times 10^{-5}\,\mathrm{m^2}$ (cf. fig. 6.2 b), $\tau_{\mathrm{dl}} = 0.003s$ is obtained, which is in the range of the measured time constants (cf. fig. 6.2 c). This hypothesis is also supported by the fact that the measured time constants correlate with the silver content – which increases the available surface area and

thus the double layer capacitance and decreases the time constant – and not with the ODC performance.
In conclusion, the semi-circles of the EIS measurement represent only the double layer charge and discharge and no mass transfer. In this context, the distorted shape of the semi-circles is surprising because the interplay of double layer capacitance and a single reaction in a homogeneous electrode would lead to a non-distorted semicircle. Although two parallel reactions pathways for the alkaline ORR may take place, both have the same rate determining step [26, 30]. Thus, no influences on the dynamics of complex kinetic processes at the electrode surface are expected and the distortion of the semi-circles is most likely not caused by two overlaying reaction processes. The distortion of the semi-circles could, however, be caused by inhomogeneous electrode properties. As studied by Lukács, a distribution of process parameters within an electrochemical system leads to detached semi-circles [90]. Paulisch et al. measured an inhomogeneous electrolyte distribution in the ODC [80]. Based on these studies, I assume that the distorted semi-circles of the ODC impedances reflect locally different reaction rates and double layer capacitances due to inhomogeneities of the electrolyte distribution. Gradients in the local reaction rates have inter alia already been reported for DMFC anodes and cathodes [91], or for acid ORR in PEMFC cathodes [92]. Based on this analysis of the experimental data, I employ a dynamic three-phase model which encompasses the effects of local inhomogeneities on the electrode to further analyze the processes in ODCs.

6.4 Modelling

It is known, that the electrodes are inhomogeneous in particle size, pore size distribution as well as in the local catalyst and binder loading [69]. These local inhomogeneities in turn lead to a distribution of hydrophilic and hydrophobic properties of the different pores [81] and may explain the previously observed electrolyte distribution [80]. The processes in the inhomogeneous three-dimensional GDE are very complex and include the effects of microkinetics [29], electrowetting [80] or closed pores [23]. Thus it is necessary to simplify the structure and processes in macroscopic model approach to describe electrode performance. The presented model focuses on the effect of an inhomogeneous electrolyte distribution and three-phase area on electrode performance. Determining an electrolyte distribution from the electrode microstructure would require a detailed understanding of factors such as pore size distributions, local surface properties inside the pores or local tortuosity, but also two-phase flow in porous media, and is out of scope of this work. The electrolyte distribution is particularly important because it governs the mass transfer paths for the water and the hydroxide ions in the liquid phase, the oxygen dissolution via the size and location of the gas-liquid interface and the area for reaction and double layer charge / discharge via the size of the wetted catalyst surface per volume.

The model of an inhomogeneous ODC is developed by dividing the ODC in multiple domains perpendicular to the through-plane coordinate z. The domains are connected in parallel and differ in the penetration depth of the electrolyte and the size of the gas-liquid interface, as shown in fig. 6.3 a). Each domain sub-model is discretised in through-plane direction (z), see fig. 6.3 b). This model is based on the assumption that the preferred mass transport is along the z coordinate in the through-plane direction, between the gas and the liquid bulk. Since the mass transport ways in in-plane direction are comparatively long (electrode thickness: 0.3 mm thick; electrode diameter: 20 mm) and run parallel to the two bulk phases, mass transport in in-plane direction is assumed to be negligible. The aim of the model is to demonstrate how the inhomogeneity influences the performance and the dynamics of the ODC. The used approach does not allow one to replicate electrode structures in all aspects, but to derive important conclusions about the impact of the inhomogeneous electrolyte distribution.

In the following, first the domain sub-model is summarized shortly. Subsequently, the interconnection of the domain sub-models to an inhomogeneous ODC is outlined. Model equations are given in table 6.1.

6.4.1 The domain sub-model

The domain sub-model is based on the expanded liquid mass transport model as introduced in chapter 5. With the help of the expanded liquid mass transport model, it was shown that – although sufficient water is available for the ORR – the dynamics and performance of the ODC are significantly affected by the mass transport of water and ions in the liquid phase. In order to consider these findings, following extensions have been made:

- As demonstrated, even under unfavorable conditions enough water is available for the ORR and does not limit the reaction kinetics in any scenario. This indicates that the common assumption is valid for ODCs, which suggest that the formation of superoxide is the rate determining step for ORR in alkaline media at a silver catalyst (cf. eq. (2.11)) [24–29]. Under consideration of the discussion in section 3.3, the reaction rate of the ORR is implemented as 1-step kinetic depending on the rate determining step, see eq. (6.6).
- It has been shown that strongly non-ideal conditions exist in the liquid electrolyte due to high ion concentration. This may affect the mass transport in the liquid phase. To consider this, Maxwell-Stefan diffusion is modeled in dependence on the gradient in chemical potential. The chemical potential of NaOH is calculated respecting the law of electroneutrality using the mean activity coefficient $\gamma_{\pm}$, (see eq. (6.21)) [32]. Fick's diffusion coefficient of oxygen is adapted by an approach given by Bird et al. [93], (see eq. (6.23)).

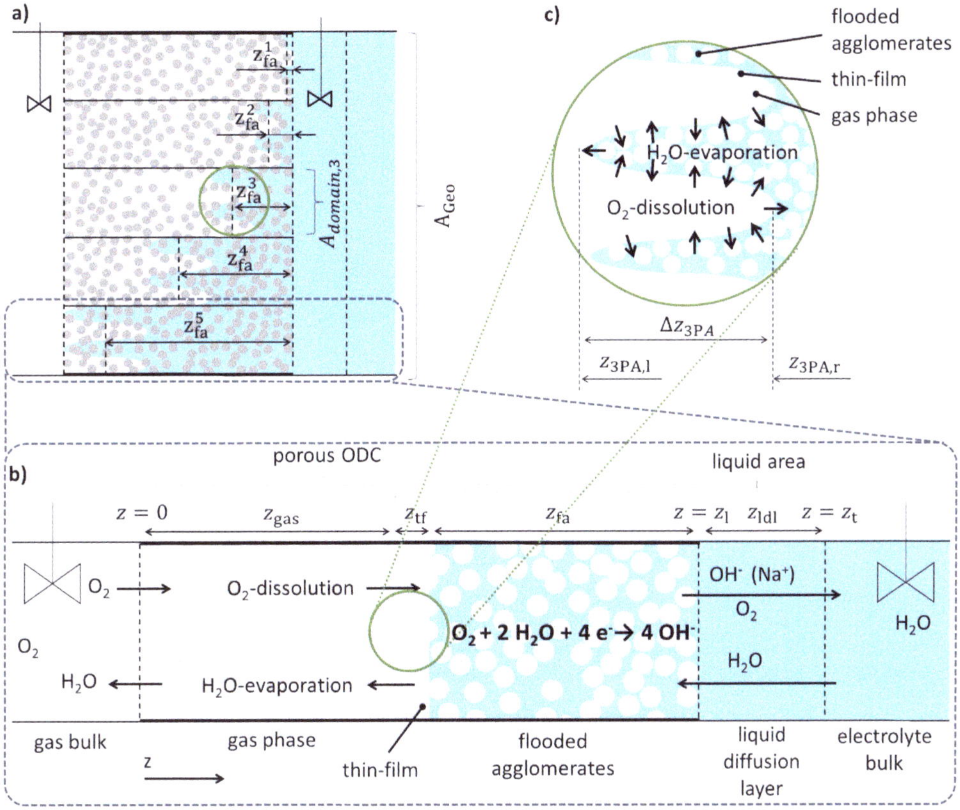

Figure 6.3: a) Schematic view of the model of an inhomogeneous ODC comprised of multiple homogeneous sub-domains; b) schematic view of the domain sub-model including reaction and mass transport; c) zoom at the gas-liquid interface.

- The calculation of the convective velocity within the liquid diffusion layer has been changed with regard to the model of an inhomogeneous ODC. It is discussed in detail in the next section.

The changes compared to the models in chapter 4 and 5 are summarized in table 3.1. All implemented model equation of the domain-sub model are given in table 6.1.

Table 6.1: Model equations for domain-sub and inhomogeneous model. (Model schema for reference: fig. 6.3.)

Electrode charge and reaction		
Faraday current	$j_F = -F \cdot 4 \cdot \int_{z_l - z_{fa}}^{z_l} r(z) dz$	(6.5)
Reaction rate [29]	$r(z) = k_{0,cat} \cdot S_{cat} \cdot a_{O_2} \cdot \exp\left(-\frac{(1-\alpha) \cdot F \cdot \eta_{reaction}}{R \cdot T}\right)$	(6.6)
Reaction overpotential	$\eta_{reaction} = E_{int} - E_0$	(6.7)
Nernst potential [29]	$E_0 = E_{00} + \frac{R \cdot T}{4 \cdot F} \ln\left(\frac{a_{O_2} \cdot a_{H_2O}^2}{a_{NaOH}^4}\right)$	(6.8)
Activity of O_2	$a_{O_2} = x_{O_2} \cdot \gamma_{O_2}$	(6.9)
Activity of NaOH	$a_{NaOH} = \frac{b_{NaOH}}{b_{NaOH}^{\Theta}} \cdot \gamma_{\pm}$	(6.10)
Molality of NaOH	$b_{NaOH} = \frac{n_{NaOH}}{m_{electrolyte}}$	(6.11)
Ohmic drop	$E_{ext} - E_{int} = R_{specific} \cdot j$	(6.12)
Charge balance	$\int_{z_l - z_{fa}}^{z_l} c_{dl,cat} \cdot S_{cat} \, dz \, \frac{dE_{int}}{dt} = j - F \cdot 4 \cdot \int_{z_l - z_{fa}}^{z_l} r(z) dz$	(6.13)
Mass transport in the gas phase and processes at the gas-liquid interface		
Species balance	$\frac{\partial p_i}{\partial t} \cdot \varepsilon = \frac{\partial^2 p_i}{\partial z^2} \cdot \mathscr{D}_{i,j}^{gas,eff} + \frac{\partial p_i}{\partial z} \cdot v_{gas}$	(6.14)
	with $i \in \{H_2O, O_2\}$	
Effective diffusion coefficient	$\mathscr{D}_{i,j}^{gas,eff} = \mathscr{D}_{i,j}^{gas} \cdot \frac{\varepsilon}{\tau}$	(6.15)
Bruggemann correlation	$\tau = \varepsilon^{-0.5}$	(6.16)
Henry's law	$c_{O_2}^* = \frac{p_{O_2}^*}{H_{O_2}(c_{NaOH}^*, c_{H_2O}^*)}$	(6.17)
Stefan flow	$\frac{p_{total}}{R \cdot T} \cdot v_{gas} = \dot{N}_{H_2O}^{evaporation} - \dot{N}_{O_2}^{dissolution}$	(6.18)
Specific phase interface area	$S_{phase,interface} = \frac{A_{phase, interface}}{A_{domain}}$	(6.19)

Mass transport in the thin film

Species balance

$$\frac{dc_i}{dt} = \sum_{\substack{j=1\\ i\neq j}}^{n} \frac{Đ_{i,j}}{RT}\left(\left(\frac{\partial^2 \mu_i}{\partial x^2} - \frac{\partial^2 \mu_j}{\partial x^2}\right)\cdot\frac{c_i c_j}{c_{\text{total}}} + \left(\frac{\partial \mu_i}{\partial x} - \frac{\partial \mu_j}{\partial x}\right)\cdot\frac{\partial c_i}{\partial x}\frac{c_j}{c_{\text{total}}} + \left(\frac{\partial \mu_i}{\partial x} - \frac{\partial \mu_j}{\partial x}\right)\cdot\frac{\partial c_j}{\partial x}\frac{c_i}{c_{\text{total}}}\right) + \frac{\partial c_i}{\partial z}\cdot v_{\text{liq,evaporation}}$$

$$\text{with } i \in \{\text{OH}^-, \text{H}_2\text{O}, \text{O}_2\},\ j \in \{\text{OH}^-, \text{H}_2\text{O}\} \tag{6.20}$$

Chemical potential of NaOH [32]

$$\mu_{\text{NaOH}} = \mu_{\text{Na}^+} + \mu_{\text{OH}^-} = \mu^{\Theta}_{\text{NaOH}} + RT\ln\left(\gamma_{\pm}^2\left(\frac{b_{\text{NaOH}}}{b^{\Theta}_{\text{NaOH}}}\right)^2\right) \tag{6.21}$$

Chemical potential of H_2O and O_2

$$\mu_i = \mu_i^{\Theta} + RT\cdot\ln(a_i) \tag{6.22}$$

$$\text{with } i \in \{\text{O}_2, \text{H}_2\text{O}\}$$

Maxwell-Stefan Diffusity [93]

$$Đ_{\text{O}_2,j} = \mathscr{D}_{\text{O}_2,j}\frac{\partial\ln(x_{\text{O}_2})}{\partial\ln(a_{\text{O}_2})} \tag{6.23}$$

$$\text{with } j \in \{\text{OH}^-, \text{H}_2\text{O}\}$$

Velocity due to evaporation

$$v_{\text{liq,evaporation}} = \frac{\dot{N}^{\text{gas,in}}_{\text{H}_2\text{O}}}{c^*_{\text{H}_2\text{O}}} \tag{6.24}$$

Mass transport in the flooded agglomerates

Species balance

$$\frac{dc_i}{dt}\cdot\varepsilon = \sum_{\substack{j=1\\ i\neq j}}^{n} \frac{Đ^{\text{liq,eff}}_{i,j}}{RT}\left(\left(\frac{\partial^2 \mu_i}{\partial x^2} - \frac{\partial^2 \mu_j}{\partial x^2}\right)\cdot\frac{c_i c_j}{c_{\text{total}}} + \left(\frac{\partial \mu_i}{\partial x} - \frac{\partial \mu_j}{\partial x}\right)\cdot\frac{\partial c_i}{\partial x}\frac{c_j}{c_{\text{total}}} + \left(\frac{\partial \mu_i}{\partial x} - \frac{\partial \mu_j}{\partial x}\right)\cdot\frac{\partial c_j}{\partial x}\frac{c_i}{c_{\text{total}}}\right) + \frac{\partial c_i}{\partial z}\cdot v_{\text{liq}} - r$$

$$\text{with } i \in \{\text{OH}^-, \text{H}_2\text{O}, \text{O}_2\},\ j \in \{\text{OH}^-, \text{H}_2\text{O}\} \tag{6.25}$$

Velocity due to reaction

$$v_{\text{liq,react}}(z) = \left(|\nu^{\text{ORR}}_{\text{H}_2\text{O}}| - |\nu^{\text{ORR}}_{\text{OH}^-}|\right)\int_{z_1 - z_{\text{fa}}}^{z_1}\frac{r(z)}{c_{\text{total}}}\text{d}z \tag{6.26}$$

Total velocity of liquid phase

$$v_{\text{liq}}(z) = v_{\text{liq,evaporation}} + v_{\text{liq,react}}(z) \tag{6.27}$$

Mass transport in the liquid diffusion layer

Species balance

$$\frac{dc_i}{dt}\cdot\varepsilon=\sum_{\substack{j=1\\i\neq j}}^{n}\frac{\text{Đ}_{i,j}^{\text{liq,eff}}}{\text{R}T}\left(\left(\frac{\partial^2\mu_i}{\partial x^2}-\frac{\partial^2\mu_j}{\partial x^2}\right)\cdot\frac{c_ic_j}{c_{\text{total}}}+\left(\frac{\partial\mu_i}{\partial x}-\frac{\partial\mu_j}{\partial x}\right)\cdot\frac{\partial c_i}{\partial x}\frac{c_j}{c_{\text{total}}}+\left(\frac{\partial\mu_i}{\partial x}-\frac{\partial\mu_j}{\partial x}\right)\cdot\frac{\partial c_j}{\partial x}\frac{c_i}{c_{\text{total}}}\right)+\frac{\partial c_i}{\partial z}\cdot v_{\text{ldl}}$$

$$\text{with } i\in\{\text{OH}^-,\text{H}_2\text{O},\text{O}_2\},\ j\in\{\text{OH}^-,\text{H}_2\text{O}\} \qquad (6.28)$$

Interconnection of the ODC domains

Specific domain area	$S_{\text{domain,k}}=\frac{A_{\text{domain,k}}}{A_{\text{geo}}}$	(6.29)
Molar flow density out of / into liquid diffusion layer	$\dot{N}^{\text{ldl}}=\sum_{k=1}^{n=5}\dot{N}_{i,k}^{\text{fa}}\cdot S_{\text{domain,k}}$	(6.30)
Molar flow density out of / into flooded agglomerates	$\dot{N}_i^{\text{fa}}=-\sum_{\substack{j=1\\i\neq j}}^{n}\left(\frac{D_{i,j}}{\text{R}T}\frac{c_ic_j}{c_{\text{total}}}\left(\frac{d\mu_i}{dx}-\frac{d\mu_j}{dx}\right)+c_i\cdot v_{\text{liq,k}}\right)\Bigg\vert_{z=z_1}$	(6.31)
Velocity	$v_{\text{ldl}}=\sum_{k=1}^{n=5}S_{\text{domain,k}}\cdot v_{\text{liq,k}}(z=z_1)$	(6.32)
Total current density	$j_{\text{ODC}}=\sum_{k=1}^{n=5}j_{\text{domain,k}}\cdot S_{\text{domain,k}}$	(6.33)

Boundary conditions

O_2 in gas bulk	$p_{\text{O}_2}(z=0)=p_{\text{total}}$	(6.34)
H_2O in gas bulk	$p_{\text{H}_2\text{O}}(z=0)=0$	(6.35)
O_2 in electrolyte bulk	$c_{\text{O}_2}(z=z_{\text{t}})=0$	(6.36)
NaOH in electrolyte bulk	$c_{\text{NaOH}}(z=z_{\text{t}})=c_{\text{NaOH}}(z=z_{\text{t}},t=0)$	(6.37)
H_2O in electrolyte bulk[a]	$c_{\text{H}_2\text{O}}(z=z_{\text{t}})=f(c_{\text{NaOH}}(z=z_{\text{t}},t=0))$	(6.38)
Electrode geometry	$z_{\text{gas}}+z_{\text{tf}}+z_{\text{fa}}=z_1$	(6.39)

[a] Calculated as in [51]

6.4.2 The inhomogeneous ODC model

This model connects the single sub-domain models to an inhomogeneously flooded ODC model. Each domain k takes up a specified proportion $S_{\text{domain},k}$ of the total geometrical electrode area

A_{geo} (eq. (6.29)). The domains are separated in through-plane direction, but connected by a common liquid diffusion layer (cf. fig. 6.3 a)). The concentrations c_i at the boundary between the flooded agglomerates and the liquid diffusion layer are obtained by solving eq. (6.30). Accordingly, the velocity of the convective flow in liquid diffusion layer is calculated by eq. (6.32). Since it has been shown that oxygen transport in the gas phase does not significantly affect the dynamics and performance of the ODC [1], it was decided not to interconnect the mass transport in the gas phase to simulate the inhomogeneities. Electrically, the domains are connected in parallel circuit, so that the total current density of the cathode equals the sum of all cathode domains, weighted by their share of the total area, eq. (6.33).

6.4.3 Properties and parameters of oxygen depolarized cathodes and representative electrode domains

In this section, the properties of the different ODCs and their representative domains are discussed. The model parameters are classified on three levels:

1. Parameters that are assumed to be identical for all ODCs and to have no distribution in ODCs: The catalyst surface area specific reaction rate constant $k_{0,\text{cat}}$ and the catalyst surface area specific double layer capacitance $C_{\text{dl,cat}}$.
2. Parameters that are assumed to be constant for each ODC, but differ between the individual ODCs: The ohmic resistance R_{specific}, the porosity ε and the specific surface area of the catalyst S_{cat}.
3. Parameters that are assumed to be distributed within the ODC: The intrusion length z_{fa} of the electrolyte and the local size of the gas-liquid interface $S_{\text{phase interface}}$ as well as relative size of the domains $S_{\text{domain,k}}$.

As it is clear that properties are distributed and that the goal of the study is to identify a distribution that can reproduce the experiments via numerical parameterization, the number of simulated domains is limited to five. This number allows to have flooded, empty and several differently flooded pores; areas with similar flooding degrees are lumped together, as the domains do only interact via the joint bulk phases. Since some parameters are independent while other parameters are connected through physical interrelationships to others, their (in)dependencies are discussed in the following.
First, an electrolyte distribution that covers a three-phase zone which extends across the entire ODC is fixed. For this purpose, the flooding length of the different domains z_{fa}, as well as the length of the three-phase areas Δz_{3PA} – calculated by their left $z_{\text{3PA,l}}$ and right boundaries $z_{\text{3PA,r}}$ –

Table 6.2: Geometrical dimensions of the three-phase area of the different domains.

Domain	$z_{\mathbf{3PA,l}}$ [a]	$z_{\mathbf{3PA,r}}$	$z_{\mathbf{fa}}$	$\Delta z_{\mathbf{3PA}}$
–	μm	μm	μm	μm
1	$0.01 \cdot z_l$	$0 \cdot z_l$	$0.005 \cdot z_l$	$0.01 \cdot z_l$
2	$0.032 \cdot z_l$	$0.01 \cdot z_l$	$0.022 \cdot z_l$	$0.022 \cdot z_l$
3	$0.1 \cdot z_l$	$0.032 \cdot z_l$	$0.066 \cdot z_l$	$0.068 \cdot z_l$
4	$0.316 \cdot z_l$	$0.1 \cdot z_l$	$0.208 \cdot z_l$	$0.216 \cdot z_l$
5	$1.0 \cdot z_l$	$0.316 \cdot z_l$	$0.658 \cdot z_l$	$0.684 \cdot z_l$

[a] Logarithmic scale from 0.01 to 1:

are considered: As illustrated in (cf. fig. 6.3 c)) the maximum intrusion depth of the electrolyte in each representative electrode domain marks the left side boundary of the three-phase area $z_{3PA,l}$. It is assumed that some pores are blocked by hydrophobic PTFE at the beginning of the electrode, while other pores are completely flooded. Thus the maximium intrusion depth of the electrolyte is assumed to be in the rang between 1 % up to 100 % of the electrode thickness. A logarithmic scale was chosen since it is expected that due to the high hydrophobicity of PTFE many pores are impermeable to the liquid electrolyte, leaving a large part of the electrode area with almost no electrolyte intrusion. In table 6.2, the values of $z_{3PA,l}$ for the five representative electrode domains are shown. The flooding degree increases monotonically from domain 1 (minimum flooding) to domain 5 (maximum flooding). A second assumption is that the three-phase area of the full electrode extends over the entire electrode thickness, but is split between the domains: For the more flooded domains, it is located closer to the gas bulk, and for the least flooded parts it is located close to the liquid diffusion layer. The right side boundary of the three-phase area $z_{3PA,r}$ of one domain equals the left side boundary of the next less flooded domain so that there are no overlaps or gaps in the three-phase area. The thickness of the three-phase area is calculated by $\Delta z_{3PA} = z_{3PA,l} - z_{3PA,r}$. The mean length of the flooded agglomerates of a domain z_{fa} is the mean value of the left side and the right side boundary of the three-phase area of the domain (cf. fig. 6.3 a), more detailed in fig. 4.1 and fig. 4.2). The oxygen gradient within the liquid electrolyte is very steep, [32, 34], thus almost no ORR outside of the three-phase area takes place, especially at higher current densities [1]. The resulting values for $z_{3PA,l}$, $z_{3PA,r}$ and z_{fa} are given in table 6.2.

Next the dependency of the local reaction rate (cf. eq. (6.6)) as well as the double layer capacitance (cf. eq. (6.13)) on the catalytically active Ag surface S_{cat} area are discussed. It can be assumed that the surface area specific reaction rate constant $k_{0,cat}$ as well as the surface area specific double layer capacitance $C_{dl,cat}$ are constant material properties. Thus differences in total catalytic activity and double layer capacitance between the different ODCs only result from differences in electrode design, and thus in the specific catalyst surface area S_{cat} and in the wetted share of the catalyst. Neumann et al. obtained an average specific surface area of

$S_{cat} = 6 \times 10^5\ m^2 m^{-3}$ Ag catalyst per electrode volume for the ODC with 97 wt−% Ag [69]. Taking these values into account, eq. (6.6) and eq. (6.13) can be used to identify $k_{0,cat}$ and $C_{dl,cat}$ from the measurements of the ODC with 97 wt−%. The values for $k_{0,cat}$ and $C_{dl,cat}$ are used for all investigated ODCs, and only the specific catalyst surface area is adjusted for each electrode. As shown before, the double layer capacitance depends on the current density [72], therefore, the value of $C_{dl,cat}$ determined here is only valid for current densities of $j = 2000\,A\,m^{-2}$.
Within ODCs the porosity ε may be distributed. Future work might elucidate how the local porosity is related to the silver/PTFE distribution and the electrolyte intrusion. Since this complex relationship is not clear yet, I use the simplifying assumption of equally distributed porosities, so that average porosity ε is set to be constant for all domains of a given electrode. The porosity, however, is dependent on the electrode design and electrode structure. It has been calculated to be 0.355 for the ODC with 97 wt−% Ag [69] and is adjusted for all further investigated electrodes and domains to reproduce the measurements.
The area-specific internal resistance of the different electrodes $R_{specific}$ was calculated from the impedance at the intersection with the real axis at high frequencies. The electric conductivity of the ODC is assumed to be high due to the high amount of Ag [32]; thus, variations in electric conductivity can be neglected and $R_{specific}$ is set to be constant for all domains. The proportion of electrode area attributed to a domain S_{domain} and the specific gas-liquid interface $S_{phase\ interface}$ of each domain have been adjusted individually to reproduce the measurements.
The model was implemented in Matlab. Measured polarization curves (cf. fig. 6.1) as well as impedance spectra (cf. fig. 6.2) have been used for manual model parameter identification. The parameters were adjusted one by one in multiple iteration loops to minimize discrepancies between experiments and simulation results. The identified parameters values for non-distributed electrode properties are given in table 6.3. The identified sizes and the specific gas-liquid interfaces for all ODC domains are given in table 6.4. All constant model parameter are given in table 6.5, all concentration-dependent parameters are given in table 6.6. The parameters and the model output are discussed in the next section.

Table 6.3: Parameters identified from (electrochemical) experiments and ODC model for electrodes with different Ag content as well as literature data.

Name	Symbol	Unit	99 wt-%	98 wt-%	97 wt-%	92 wt-%
Reaction rate constant	$k_{0,cat}$	$mol\,m^{-2}\,s^{-1}$	0.39			
Double layer capacitance	$C_{dl,cat}$	$F\,m^{-2}$	1.22			
Porosity	ε	–	0.4	0.37	0.355^a	0.26
Specific catalyst interface	S_{cat}	$\mu m^2\,\mu m^{-3}$	0.92	0.79	0.6^a	0.58
Specific resistance[b]	$R_{specific}$	$\Omega\,m^2$	4.15×10^{-5}	2.83×10^{-5}	2.32×10^{-5}	2.82×10^{-5}

[a] Determined by [69]. [b] High frequency resistance from EIS measurements.

Table 6.4: Distributed parameters identified from electrochemical experiments and ODC model for electrodes with different Ag content.

Name	Symbol	Unit	Domain	99 wt-%	98 wt-%	97 wt-%	92 wt-%
Specific domain area	S_{domain}	%	1	3.20	5.19	12.1	22.0
			2	10.4	16.86	16.9	21.0
			3	24.0	37.74	38.6	35.2
			4	18.4	22.05	19.3	12.3
			5	44.0	18.16	13.3	8.5
Specific gas-liquid interface per thickness of three-phase area	$\frac{S_{phase\ interface}}{\Delta z_{3PA}}$	$\mu m^2\,\mu m^{-3}$	1	2.94	4.48	3.54	2.68
			2	8.71	9.89	7.65	6.02
			3	4.30	6.55	3.44	1.90
			4	0.08	0.15	0.11	0.05
			5	0.02	0.03	0.02	0.01

Table 6.5: List of constant parameters and operating conditions.

Name	Symbol	Unit	Value
Operating conditions			
Pressure in gas chamber[a]	$p = p_{O_2}$	Pa	1.01×10^5
NaOH concentration in electrolyte bulk[a]	$c_{NaOH}(z = z_t)$	$mol\,m^{-3}$	9.68×10^3
H_2O concentration in electrolyte bulk [51]	$c_{H_2O}(z = z_t)$	$mol\,m^{-3}$	50.0×10^3
Temperature[a]	T	K	353.15
Electrochemical and kinetic data			
Charge transfer coefficient[b] [24]	α	–	0.15
Stoichiometric coefficient H_2O	$\nu_{H_2O}^{ORR}$	–	−2
Stoichiometric coefficient OH^-	$\nu_{OH^-}^{ORR}$	–	+4
Open circuit potential [23]	E^0	V vs. RHE	1.13
Binary diffusion coefficients			
For gaseous oxygen/water [73]	$\mathscr{D}_{O_2,H_2O}^{gas}$	$m^2\,s^{-1}$	2.95×10^{-5}
For liquid oxygen/water[c] [73, 74]	$\mathscr{D}_{O_2,H_2O}^{liq}$	$m^2\,s^{-1}$	2.18×10^{-9}
For liquid oxygen/NaOH[c] [73, 74]	$\mathscr{D}_{O_2,NaOH}^{liq}$	$m^2\,s^{-1}$	3.25×10^{-9}
For liquid water/NaOH[d] [60]	$Đ_{H_2O,NaOH}^{liq}$	$m^2\,s^{-1}$	1.26×10^{-9}
Geometrical parameter			
Thin-film thickness [32]	z_{tf}	m	60×10^{-9}
Thickness of liquid diffusion layer[e] [2]	z_{ldl}	m	10×10^{-6}

[a] Experimental operating condition. [b] Measured with 6.5 M NaOH. [c] Calculated with modified Wilke and Chang equation, with $\Theta = 3.9$ and $a = 0.5$ chosen, adjusted to experimental data from [74]. [d] No NaOH data available, modeled for KOH with NaOH viscosity from [68], described in [32]. [e] Estimated to be independent of current density. Taken from my previous study where it was identified to be $10\,\mu m$ for measurements with vertically orientated ODC [2], the measurement results in the set-up used here correspond to those in the reference cell [23].

6.5 Comparative model-based analysis of inhomogeneous oxygen depolarized cathodes

In fig. 6.2 a) the experimental and simulated impedance spectra are presented. The simulations reproduce the measurements well regarding the distorted shape of the semi-circles, the trend in charge transfer resistance with silver content (fig. 6.2 b)), and the time constant (fig. 6.2 c)). Additionally, in fig. 6.1, the experimental and simulated polarization curves are given, which also agree with slight deviations in the range of very high and very low current densities. Deviations in the low current range may result from the change of the Tafel-slop at a overpotential of about 0.16 V vs. OCP (IR-corrected) [23, 32]. As discussed in [32], the reason for this behavior is not clear, possible explanations may be different adsorption isotherms of intermediates in low the

Table 6.6: Electrolyte concentration-dependent variables.

Name	**Symbol**	**Unit**
Inverse Henry constant[a] [22]	H_{O_2}	$Pa\,m^3\,mol^{-1}$
Activity coefficient of O_2[b] [50]	γ_{O_2}	–
Activity of H_2O[c] [75]	a_{H_2O}	–
Mean activity coefficient of NaOH [66]	$\gamma_{\pm}$	$kg\,mol^{-1}$
Density of electrolyte [68]	ρ	$kg\,m^{-3}$
Water vapor pressure [51]	$p_{H_2O}^{vap}$	Pa

[a] Extrapolated from $c_{NaOH} \leq 6M$, but as shown in [27] extrapolated values fit very well to experiments with $c_{NaOH} \leq 12M$, [b] Extrapolated from $c_{NaOH} \leq 5.5M$, but as shown in original research, calculations for e.g. $c_{KOH} \leq 13.5M$ are valid [50], [c] Extrapolated data from 343.15 to 353.15 K.

potential region [94], or influences by electronic or ionic conductivities which also could lead to a change in Tafel-slope [95]. Even though the experimental setup is temperature-controlled [23], the electrode may heat up at high currents. This could explain the deviations in the higher current range. Moreover, it is possible that the inhomogeneities are not limited to an inhomogeneous electrolyte distribution. This may lead to effects not considered in the model. The model is additional validated against EIS spectra of the high performing ODC with 98 wt−% Ag at current densities of $1\,kA\,m^{-2}$ and $4\,kA\,m^{-2}$. As shown in the appendix D (fig. D.1) the simulation data agree well with the experimental impedance spectra in term of shape, time constant and charge transfer resistance. All in all, the presented model can describe the electrode behaviour reasonably well.

In the following, first the influence of inhomogeneities on electrode performance will be discussed by analysing the contributions of the representative electrode domains to the polarization curves. Subsequently, the influence of the inhomogeneities on the dynamic behaviour will be discussed by analysing the EIS spectra. The parameters of the individual ODCs in table 6.3 and 6.4 are evaluated and compared in more detail during the analysis.

In fig. 6.4, the individual contributions of the different domains towards the overall performance are shown for the ODCs with 92 and 98 wt−% Ag.

The local current density of each domain is plotted versus the potential. In both ODCs the representative domains strongly differ in their performance. For the ODC with 92 wt−% Ag, the current density in domain 1, which has the lowest electrolyte intrusion, approaches a plateau at around 0.5 V, indicating a depletion of oxygen in the liquid phase. In contrast, the medium flooded domains 2 and 3, amounting to ca. 50 % of total electrode area, are most active and have not yet reached their limiting current density at 0.5 V vs. RHE. The highly flooded domains 4 and 5, amounting to ca. 40 % of electrode area, contribute only little to the total current density. Even though the ODC with 98 wt−% has comparable proportions of active regions, it reaches a significantly higher performance. To understand this discrepancy in performance between the

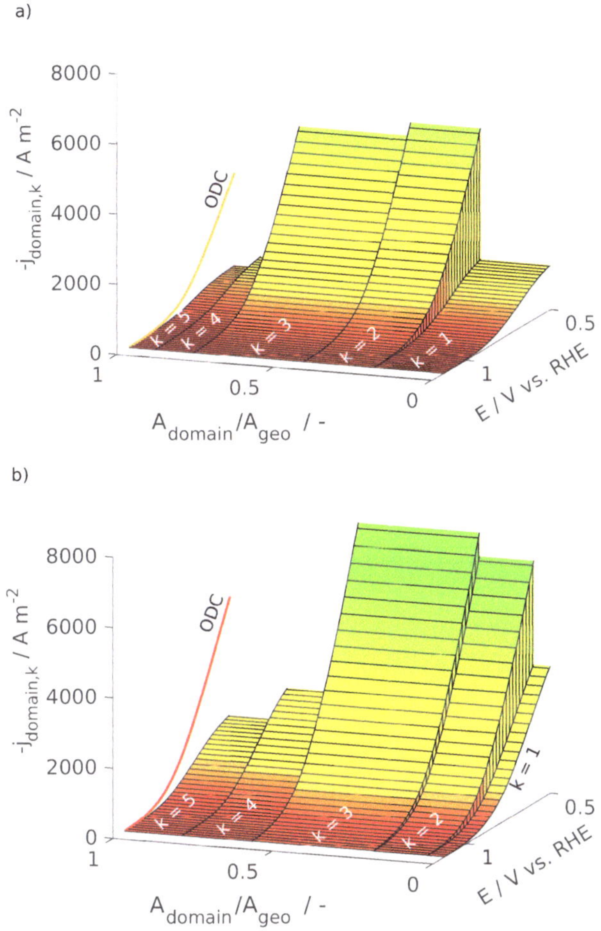

Figure 6.4: Simulated current density dependent on the potential in each domain k and for total ODC, for ODCs with a) 92 wt−% Ag and b) 98 wt−% Ag.

two ODCs and their different domains, the differences of electrolyte intrusion as well as the gas-liquid interface have to be taken into account. In fig. 6.5 a), the proportion of the pore volume that is filled with electrolyte, p_v, is displayed over the dimensionless electrode thickness for all ODCs. The proportion is calculated according to fig. 6.3 a) by accumulating the relative sizes of the domains k, which is equal to the relative volume fraction of the respective domains, that are flooded at a respective location. It is assumed that pores are flooded to different degrees in the three-phase zone. For the calculation an average degree of 50 % flooding was used, since no further information on the degree of flooding in the three-phase area is available. Thus, only half of the relative size of the domain is taken into account if the three-phase area is present at the respective location. Eq. (6.40) describes the calculation of the electrolyte filling over the electrode depth, z_d is the distance from the end of the electrode facing the liquid diffusion layer:

$$p_\mathrm{v}(z_\mathrm{d}) = \begin{cases} \sum_{k=1}^{1} S_\mathrm{domain,k} - 0.5 \cdot S_\mathrm{domain,1} & \text{for } z_\mathrm{d} < 0.01 \cdot z_\mathrm{l} \\ \sum_{k=1}^{2} S_\mathrm{domain,k} - 0.5 \cdot S_\mathrm{domain,2} & \text{for } 0.01 \cdot z_\mathrm{l} \leq z_\mathrm{d} < 0.032 \cdot z_\mathrm{l} \\ \dots & \dots \\ \sum_{k=1}^{5} S_\mathrm{domain,k} - 0.5 \cdot S_\mathrm{domain,5} & \text{for } 0.316 \cdot z_\mathrm{l} \leq z_\mathrm{d} < z_\mathrm{l} \end{cases} \tag{6.40}$$

Among the electrodes a trend of decreasing electrolyte saturation with higher PTFE content is clearly visible. This agrees with the results of a radiographic study of Paulisch et al. on similar electrodes, who found that a higher Ag content leads to a higher electrolyte saturation [80]. The effect can be attributed to the hydrophobic properties of the PTFE. According to the simulation results, also the porosity of the ODCs decreases with increasing PTFE content – this trend agrees qualitatively with FIB/SEM analyses of ODCs with varying Ag amount [23]. The porosity change may have an additional effect on the wettability of the electrodes as smaller pore diameters cause higher capillary forces. Although the wetted catalyst surface is the largest in the 99 wt−% Ag, this ODC has the lowest performance. This is interpreted as an indication that the actual electrochemical reaction rate is not the performance limiting step, even at medium current densities. The poor solubility of oxygen in liquid electrolytes is a well-known phenomenon [22]. Therefore the size of the gas-liquid interface can limit the performance. With the simulation data, an estimation of the size of the interface dependent on the location within the ODCs has been made, which is given in fig. 6.5 b). The specific area of the gas-liquid interface within a

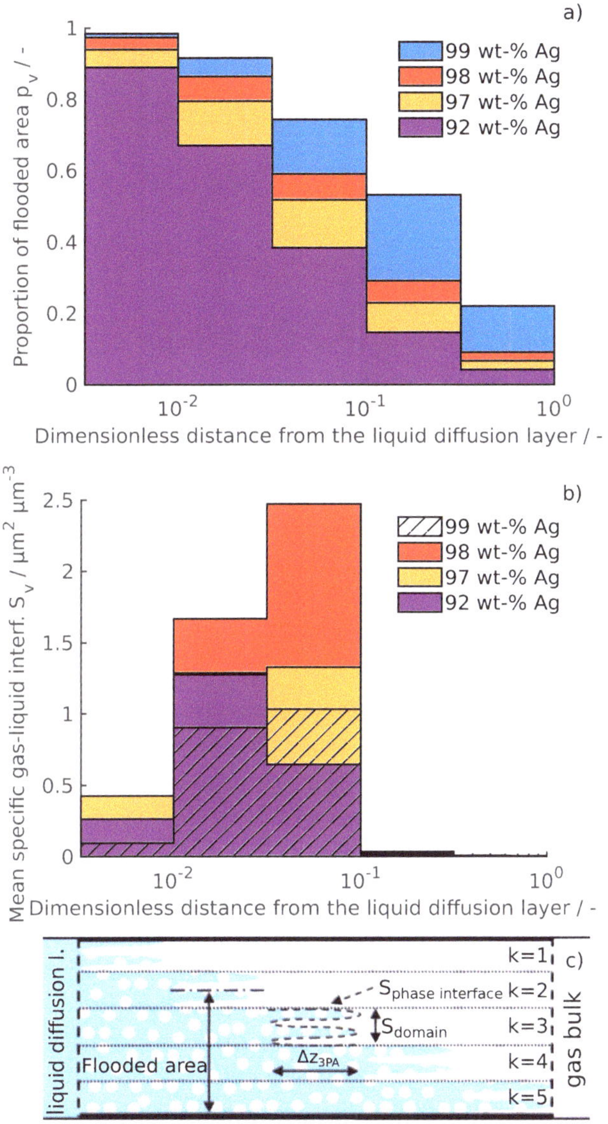

Figure 6.5: Identified properties of ODCs with different Ag-content: a) Electrolyte distribution dependent on the dimensionless location within the gas diffusion electrode (calculated by eq. (6.40)); b) average area of gas-liquid interface per electrode volume dependent on the dimensionless location within the gas diffusion electrode (calculated by eq. (6.41)); c) Schematic illustration of the electrolyte intrusion and gas-liquid interface distribution for reference.

given volume at each location z_d is calculated by scaling the specific gas-liquid interfaces of the respective domains according to their share of the total electrode area (eq. 6.41):

$$S_\mathrm{v}(z_\mathrm{d}) = \begin{cases} \frac{S_\text{phase interface,1}}{\Delta z_\mathrm{3PA,1}} \cdot S_\text{domain,1} & \text{for } z_\mathrm{d} < 0.01 \cdot z_\mathrm{l} \\ \frac{S_\text{phase interface,2}}{\Delta z_\mathrm{3PA,2}} \cdot S_\text{domain,2} & \text{for } 0.01 \cdot z_\mathrm{l} \leq z_\mathrm{d} < 0.032 \cdot z_\mathrm{l} \\ \dots & \dots \\ \frac{S_\text{phase interface,5}}{\Delta z_\mathrm{3PA,3}} \cdot S_\text{domain,5} & \text{for } 0.316 \cdot z_\mathrm{l} \leq z_\mathrm{d} < z_\mathrm{l} \end{cases} \tag{6.41}$$

According to fig. 6.5 b), the ODC with 98 wt−% Ag exhibits the largest gas-liquid interface and – as the reaction mostly takes place close to the phase interface – this electrode has thus the largest active three-phase region and the lowest charge transfer resistance. This causes the high current densities presented in fig. 6.4 b). It seems that the electrode composition of 98 wt−% Ag and 2 wt−% PTFE leads to an optimal balance between hydrophilic and hydrophobic properties and thus the highest performance. In contrast, the ODC with 99 wt−% Ag is highly flooded so that the low performance can be attributed to missing gas phase. On the other side, the ODC with 92 wt−% Ag is barely flooded. Thus the gas-liquid interface is too small due to insufficient electrolyte intrusion. Furthermore, the model suggests what would be intuitively expected: Stronger hydrophobic properties of the electrodes lead to a shift of the three-phase boundaries towards the gas side. Almost the full gas-liquid interface, and thus active area, is located within the first 10 % of the electrode from the liquid side whereas much of the Ag electrode at the gas side is not electrochemically active. This also suggests that there is mostly a small distance only from the gas-liquid interface to the liquid bulk, and thus little mass transport resistance, which agrees with previous studies [2].

The claim of negligible mass transport limitation is also supported by the time constant analysis of the EIS measurements in section 2. As presented in a previous chapter, long mass transport ways in the liquid phase between the electrochemical active zone and the electrolyte reservoir lead to characteristic time constants [2]. Due to the fact that no mass transport time constants in any ODC were measured and due to the fact that the most flooded electrode with 99 wt−% Ag performs worst, it is concluded that the flooded domains 4 and 5 of the ODCs contribute very little to the total electrochemical activity. The conclusion that the highly flooded domains are inactive is counter intuitive, since in this domains both the gas phase and the liquid phase are present, thus the presence of a physical gas-liquid interface is expected (cf. fig. 6.3). The electrochemical measurements as well as the macroscopic model strongly suggest that the three-phase areas in the middle of the ODCs and close to the gas side are electrochemically almost "dead". To clarify the mechanism, further simulations on a pore scale or detailed experimental studies that link the electrode microstructure with the local performance would be necessary. Possible reasons for the low activity of largely flooded pores might be long mass transport ways across the ODCs or bottlenecks in the pore system leading to a high ionic resistance. Perhaps also local salt accumulations due to the ORR [34] lead to a quasi insolubility of oxygen. The availability of oxygen affects the OCP and the reaction rate of the ORR. In fact, the here presented gas-liquid interface should be interpreted as the active interface of ionically and electrically well-connected pores with exposed silver surface, and not the physical gas-liquid interface.

Now the impact of inhomogeneities on EIS is discussed. Using the presented model, the different local electrolyte saturation and the differences in local performance are shown to be responsible for the distorted shape of the semi-circles in the impedance spectra. The sinusoidal current of the total ODC is composed of the sinusoidal currents of all single domains. It represents the time constants and thus the phase shifts of all domains. The domains differ in double layer capacitance, in local performance, and thus in charge transfer resistance, therefore they all are characterized by their own time constant (cf. eq. (6.4)). The impedance spectra can be examined in more detail by including the Bode plots, which are given for the example of the ODC with 98 wt−% Ag in fig. 6.6 a) and b). Besides the phase shift φ and impedance $|Z|$ of the entire ODC, the impedance values for the individual domains are also plotted separately.

The minima in phase shift of the domains extend over four orders of magnitude in frequency as shown in fig. 6.6 a). The phase shift φ at higher frequencies is governed by the less flooded domains (domains 1 and 2) and the more flooded domains (domain 4 and 5) govern the lower frequency range. The resulting overall phase shift curve reproduces the measurements well.

The model also is able to reproduce the trend of the EIS time constants over the Ag content which is evident in the experimental results. In fig. 6.2 b), it can be seen that the characteristic time constant is slightly increasing with the silver content in experiment and in simulation. The characteristic time constant equals the product of the double layer capacitance and the charge transfer resistance (cf. eq. (6.4)). While R_{ct} from 92 to 99 wt−% Ag only increases by about a factor of 1.5, the time constant increases by a factor of 20 (cf. fig. 6.2). A major part of the increase in the time constant can be attributed to an increase in double layer capacitance: Due to the more hydrophilic properties of the ODCs with a higher Ag ratio, a higher part of the ODCs is flooded, which leads to a larger share of wetted catalyst surface, where the double layer charge and discharge takes place. Additionally, the higher amount of silver catalyst within the ODCs also results in a higher available total Ag surface area S_{cat}. This also explains, why the time constant of the ODC with 92 wt−% Ag is faster than the time constant of the ODC with 98 wt−% Ag although the charge transfer resistance increases. The identified double layer capacitances are within the range of previous studies [72].

As mentioned before, a slightly different shape of the EIS spectrum was observed for ODC with 99 wt−% Ag: The impedance spectrum shows a second, less pronounced semi-circle. According to the model results, the higher frequency semicircle can be assigned to the active pores, while the lower frequency semicircle is assigned to the flooded pores. In the remaining ODCs, the proportion of flooded pores is lower, so the impedance spectra of the flooded parts are less significant in the overall impedance spectra.

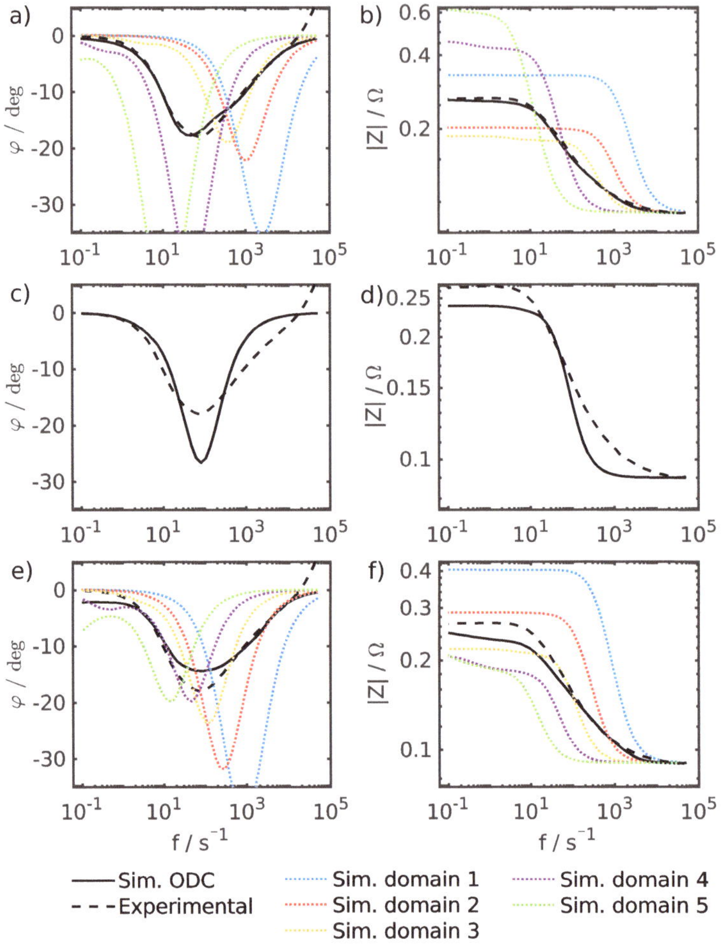

Figure 6.6: Bode plots for EIS at $j = 2000\,\mathrm{A\,m^{-2}}$ for ODC with 98 wt−% Ag for full electrode and single electrode domains. The three scenarios: a), b) inhomogeneous ODC (five domains); c), d) homogeneous ODC (one domain; $\frac{S_{\mathrm{phase\ interface}}}{\Delta z_{\mathrm{3PA}}} = 1.06\,\mu\mathrm{m}^2\,\mu\mathrm{m}^{-3}$); e), f) semi-homogeneous ODC (five domains, $\frac{S_{\mathrm{phase\ interface}}}{\Delta z_{\mathrm{3PA}}} = 6.4\,\mu\mathrm{m}^2\,\mu\mathrm{m}^{-3}$; $S^k_{\mathrm{domain}} = 20\,\%$).

6.6 Model-based scenario analysis of impedance spectra

As shown so far, the combination of dynamic and steady state electrochemical measurements and simulations enables to detect and quantify inhomogeneities and the related parameters in ODCs; the influence of electrolyte distribution within gas diffusion electrodes on their electrochemical performance and dynamic behaviour is clearly visible. A methodology which can identify and quantify inhomogeneities is a highly promising tool that enables knowledge-driven, targeted design of electrodes and state diagnosis. To get deeper insights into effects of the parameter distributions and to evaluate the impedance spectra and polarization curves for their informative value and diagnostic capability regarding the electrolyte distribution, a scenario analysis is presented in the following: Two hypothetical alternative model scenarios are analyzed for the high-performing ODC with 98 wt−% Ag, the results are compared with the above presented experimentally parameterized model for an ODC with 98 wt−% Ag. In the first scenario, a completely homogeneous ODC with only one domain is modeled: Here no influences due to an inhomogeneous electrolyte distribution are expected. The second scenario is an ODC with five domains with different degrees of flooding, where the flooding length z_{fa}^{k} is identical to that given in table 6.2; however, here equidistant domain areas and a constant specific gas-liquid interface over the whole electrode are assumed: $\frac{S_{\text{phase interface}}}{\Delta z_{\text{3PA}}} = 6.4\,\mu\text{m}^2\,\mu\text{m}^{-3}$ and $S_{\text{domain}}^{k} = 20\,\%$. The gas-liquid interface has the same surface area per electrode volume at each position in z-direction within the ODC, but the electrode maintains its different degrees of flooding. The model parameters for the two hypothetical scenarios are adjusted manually to best reproduce the polarization curve as well as the charge transfer resistance and time constant of the EIS. Model parameters and model characteristics are given in the appendix D.

The outcome of the scenario analysis is summarized in fig. 6.7 where the simulated EIS spectra and polarization curves for all three scenarios are shown. All three model variations can reproduce the polarization curves, only minor differences between the model outputs are visible. However, in case of impedances, the simulated inhomogeneous ODC model matches the EIS spectra most accurately. In contrast, the homogeneous ODC shows a single non-distorted semicircle, while the ODC with semi-homogeneous properties exhibits a significant smaller impedance over the frequency range up to 1 kHz, i.e. it has a lower charge transfer resistance. In addition, the model of a semi-homogeneous ODC shows signs of mass transport impacts, visible in the additional arc visible at low frequencies below 1 Hz.

The underlying effects can be explained on the basis of the Bode diagrams. In fig. 6.6 c) and d) the Bode plot for the homogeneous ODC is illustrated. It can be seen that the minimum in phase shift is spread over approximately two orders of magnitudes in frequency. The phase shift curve is clearly not as wide as the measured phase shift, since a distribution of time constants is missing. The semi-homogeneous ODC features a phase shift over a wide frequency range (fig.

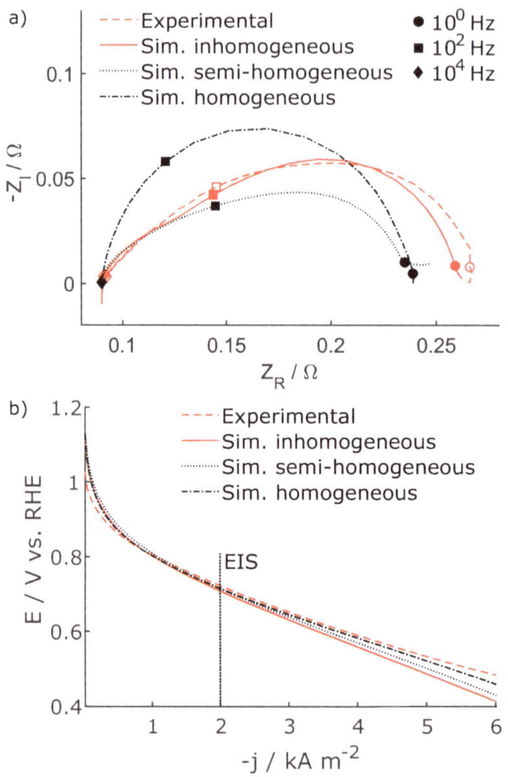

Figure 6.7: Experimental and different simulated scenarios: a) EIS at $j = 2000\,\mathrm{A\,m^{-2}}$ for ODC with 98 wt−% Ag; b) polarization curve.

6.6 e)). However, its phase shift as well as the impedance (fig. 6.6 f)) of the semi-homogeneous ODC diverge from the measurements at frequencies below 1 Hz, i.e. where mass transport effects would be expected. It can be seen from the lower magnitude of the impedance $|Z|$ that the highly flooded domains both 4 and 5 exhibits a small $|Z|$ and thus contribute significantly to the total current. In these domains, the mass transport length of the liquid electrolyte within the ODC is so long that a time constant of the water and hydroxide ion mass transport at frequencies below 1 Hz can be observed. This results in overlapping mass transport semi-circles of the highly flooded domains 4 and 5 that form the peculiar almost horizontal line in the Nyquist diagram. Please note that due to the defined diffusion length no Warburg-type behavior is expected. As previously discussed, as a mass-transport related behavior is not visible in the experimental data, it can be concluded that such a semi-homogeneous condition is not present in the experimentally analyzed electrodes.

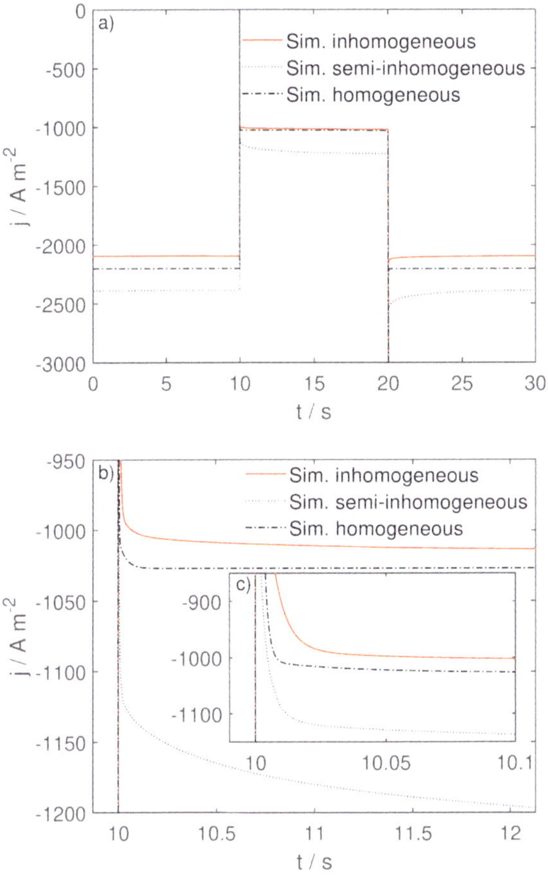

Figure 6.8: Different simulated scenarios: a) Time-dependent current density response for a potential step from 0.8 to 0.7 V vs. RHE at $t = 10$ s and back at $t = 20$ s; b) and c): Zooms of a).

Methodologically, the question arises whether the information about the inhomogeneities in the electrolyte distribution is also accessible by means of potential steps as used in the chapters 4 and 5. To evaluate this, potential steps from 0.8 to 0.7 V vs. RHE and back were simulated with the parameter sets of the three scenarios. The resulting time-dependent current densities are shown in figure 6.8 a), zooms are given in fig. 6.8 b) and 6.8 c).

As shown in figure 6.8 a), all three scenarios have a positive current peak immediately after the positive potential step at 10 s and a negative peak immediately after the negative potential step at 20 s, which are indicators for a fast process; for the semi-inhomogeneous a pronounced and for the inhomogeneous ODC a less pronounced slower relaxation follows the peaks, which are indicators for a slow process. The different current densities of the three scenarios after relaxation result from the different polarization behavior of the scenarios, cf. fig. 6.7 b). Similarities and

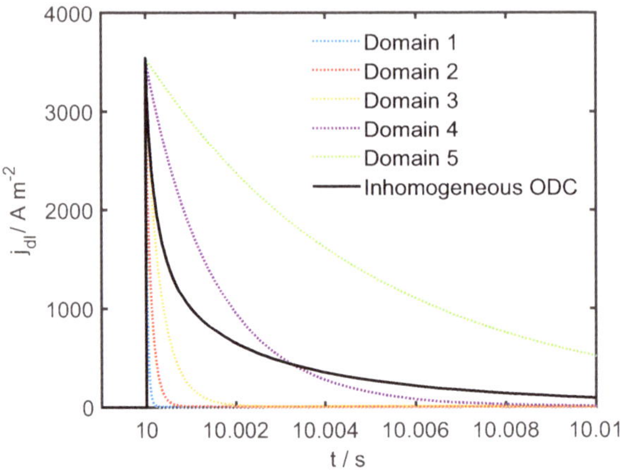

Figure 6.9: Time-dependent capacitiv current density response for a potential step from 0.8 to 0.7 V vs. RHE at $t = 10\,\mathrm{s}$ for full electrode and single electrode domains of the inhomogeneous ODC.

differences in the dynamic behavior of the three scenarios are discussed in the following.
The peaks are caused by the double layer charge and discharge as the fast process. Only with a strong zoom on the positive current peaks (fig. 6.8 b) and c)), differences between the double layer charge currents of the scenarios are recognizable. The peaks of the inhomogeneous and the semi-homogeneous ODC are widened compared to the peak of the homogeneous ODC, which is a result of the parameter distribution. To investigate this more deeply, the capacitive double layer charge current densities j_{dl} of the inhomogeneous ODC and its domains are given in fig. 6.9. As can be seen, the width of the double layer peaks and thus the time constant of the domains increases with their flooding degree. The double layer charge current of the full electrode is the sum of the charge currents of its domains, weighted by their share of the total area, cf. eq. (6.33). In the interval $10\,\mathrm{s} < t \leq 10.001\,\mathrm{s}$, a pronounced decrease in double layer charge current of the ODC can be noticed, which is attributable to the fast time constants of the less flooded domains 1 to 3. But there is still charging at $10.001\,\mathrm{s} < t \leq 10.01\,\mathrm{s}$, which is caused by the comparable long time constants of the strongly flooded domains 4 and 5. Concluding, the contributions of the strongly flooded domains 4 and 5 are mainly responsible for the widen shape of the double layer peak of semi-homogeneous and the inhomogeneous ODC. Due to the complex shape of the peaks, a determination of the double layer charge time constant based on the current response of the applied potential steps is not possible.

After the double layer peaks, a pronounced, slower relaxation can be noticed for the semi-homogeneous ODC. The same effect is also weakly emerging for the inhomogeneous ODC, but not for the homogeneous ODC. The underlying dynamic process is the mass transport of water and hydroxide ions in the liquid phase between the reaction zone and the liquid bulk (see chapter 5). Two characteristics must be considered when comparing the relaxation behavior of the

current densities: On the one hand it is the time constant of the process, on the other hand it is the magnitude of the change in current density during the relaxation. The magnitudes of the relaxation amounts about 180 $A\,m^{-2}$ for the semi-inhomogeneous ODC, but only about 10 $A\,m^{-2}$ for the inhomogeneous ODC. Accordingly, the mass transport of water and ions in the liquid phase has a large influence on the dynamics of the semi-inhomogeneous ODC, but only a very limited influence on the dynamics of the inhomogeneous ODC. As discussed in the impedance analysis, the more significant mass transport in the semi-inhomogeneous is caused by the larger contribution of domains 4 and 5 to the total current density. In the homogeneous ODC, the electrode dynamics is only marginally affected by the mass transport, therefore this process is not pronounced at all. Analogous to the double layer peak, it is not possible to determine the mass transport time constant for ODCs with distributed parameters by using potential steps.
In summary, the simulations of the potential steps enable to reveal the double layer charging and the mass transport influences. Although the information about the distribution in double layer capacitance and charge transfer resistance is reflected in the current responses, they are difficult to identify, interpret and especially to quantify. As discussed previously, this information is accessible with impedance analysis. Thus, EIS seems to be a more suitable tool for the analysis of these processes. For the inhomogeneous ODC, a minimal relaxation due to the mass transport of water and ions in the liquid phase was detected, which remained hidden in the EIS spectra. It seems that the mass transport is slightly better resolved by potential steps than by EIS, although the time constant of this process cannot be determined by the potential step analysis. However, since the current changes by only about 1 % during relaxation, the potential steps indicate that the mass transport in the liquid phase is of minor importance for the dynamics of the inhomogeneous ODC.

Concluding, although all scenarios match the steady state data, only the inhomogeneous model can reproduce the impedance spectra. The information about the inhomogeneities is difficult to access by means of potential steps. In the impedance spectra, the homogeneous electrolyte distribution does not meet the distorted semi-circles; the semi-homogeneous distribution of the electrolyte exhibits an additional time constant due to the active regions close the end of the ODC facing the gas side. The ability of EIS for detecting time constant and their distributions implies that it is a suitable tool for investigating and quantifying the gas-liquid distribution in gas diffusion electrodes and its impact on the electrode performance.

6.7 Conclusions

In this chapter I investigated the influence of electrolyte distribution for ODCs with different binder-silver ratios on electrode performance and electrode dynamics using experimental as well

as modeled polarization curves and impedance spectra.
Experimental impedance spectra of oxygen depolarized cathodes with different Ag- and PTFE-content have been presented which were mainly characterized by fast time constants of $\tau \leq 0.03\,\mathrm{s}$ and distorted semi-circles in the Nyquist plot. For a deeper understanding of the impedance spectra, the first model of an ODC with distributed properties was presented. The fast time constants can be assigned to the double layer charge and discharge, whereas the distorted semi-circles are a result of locally varying electrolyte intrusion depth. The model results suggest that pore systems with a low electrolyte intrusion make a significantly higher contribution to the total performance than the highly flooded pores. A comparative model-based analysis of ODCs with different silver to PTFE ratios has shown that the electrodes differ significantly in performance and dynamics. Performance is not attributed to a large wetted catalyst surface but to a large active gas-liquid interface that is located close to the electrolyte bulk. Furthermore, the model results were compared with two hypothetical scenarios of a homogeneous ODC and an ODC which is active over the entire electrode thickness. It was shown that the homogeneous ODC exhibits undistorted semi-circles, whereas the electrode with the semi-homogeneous gas-liquid interface shows mass transport effects, which are both not visible in the experimental impedance spectra. The presented insight will aid experimentalists and electrode designers to better diagnose the state of their electrode using EIS and to judge how heterogeneous or distributed the electrolyte phase is. EIS will also help to monitor electrodes: As electrodes age, their hydrophobicity might change and thus electrolyte distribution, EIS and performance. Future studies are needed to investigate the local performance and the structure on a pore-scale level and to establish a direct link between the microstructure and ODC performance.

7 Conclusions and Perspective

In this chapter, the dissertation is summarized and key findings are highlighted. Additionally, perspectives and open research questions regarding ODCs are discussed.

7.1 Summary and conclusions

The goal of this dissertation is to gain in-depth understanding of the processes in ODCs for advanced chlor-alkali electrolysis in order to identify limitations and to provide the fundamentals for a knowledge-based electrode optimization. To achieve this goal, dynamic three-phase models were developed, parameterized and systematically improved based on simulation results and current literature findings. The novelties of the models are in particular (i) the dynamic approach, (ii) the appropriate consideration of the water activity and water mass transport, (iii) the inclusion of the phase equilibrium dependent on the ion and water concentration as well as (iv) the representation of structural inhomogeneities of the porous ODCs.

Initially, a basic model of an ODC was developed to investigate the processes and limitations in ODCs. With the help of the basic model it was demonstrated that under industrial operating conditions ORR only takes place close to the gas-liquid interface due to a lack of available oxygen in the liquid phase. A key factor for the low availability of oxygen in the liquid electrolyte is its poor solubility as a result of high ion concentration and low water activity. Since during ORR water is consumed and hydroxide ions continue to accumulate near the gas-liquid interface, the oxygen solubility decreases further with increasing current density. This effect contributes significantly to the total depletion of oxygen within the liquid electrolyte at high reaction rates and thus to a limitation in current density. A dynamic analysis was performed, where the mass transport of the water and hydroxide ions in the liquid phase was identified as the slowest process in the ODC. Since water and ion concentrations affect the oxygen solubility as well as the water evaporation, the dynamics of concentration or partial pressure profiles of all species follow this process.

For an in-depth analysis of water and ion mass transport in the liquid phase and its influence on the ODC performance, the expanded liquid mass transport model was developed. With the help

of this model, electrochemical measurements of two setups[11], differing regarding the water and ion mass transport in the liquid phase, were compared. The model-based analysis of dynamic and steady-state measurements has shown that a forced convective mass transport in the liquid phase accelerates the dynamics and increases the performance of an ODC compared to an ODC without a forced convection within the liquid electrolyte. The model suggested that the forced convective flow reduces the thickness of the liquid diffusion layer by a magnitude, which leads to faster the mass transport in the liquid phase. As a result the accumulation of hydroxide ions and thus the decreasing water activity and oxygen solubility is less pronounced. As demonstrated, water activity does hardly influence the kinetics of the ORR directly, but rather indirectly via the oxygen solubility.
Next, it was investigated how the structure of an ODC and an inhomogeneous electrolyte distribution within the ODC affect the water and ion mass transport, the local oxygen availability and finally the electrode performance. For this purpose, the inhomogeneous model was developed. Using the inhomogeneous model, polarization curves and EIS measurements[12] of ODCs with different binder-catalyst ratios were investigated. It was demonstrated that the EIS measurements of all ODCs are characterized by fast time constants of $\tau \leq 0.03\,\mathrm{s}$ and distorted semi-circles in the Nyquist plot. While the fast time constants can be assigned to the double layer charging and discharging, the distorted semi-circles seem to be a result of inhomogeneous electrolyte distributions within the ODCs. The model results suggested, that electrode areas with a high electrolyte saturation are almost electrochemically inactive and barely contribute to the electrode performance. The dynamics and performance of the ODCs with different binder-catalyst ratio were compared with aid of the inhomogeneous model. It was shown that not the area of the wetted catalyst, but the size and location of the gas-liquid interface within the ODCs are the key driver to a high electrode performance.

Concluding this work, the understanding of processes and limitations within ODCs as well as their dependency on the environmental conditions and electrode structure was expanded fundamentally. These insights are relevant for both academia and industry. It was shown that an enrichment of ions at the gas-liquid interface and the corresponding decrease in water activity limits oxygen availability in the liquid phase and thus ODC performance. As concluded, the performance of technical ODCs can be optimized by a higher mass transfer rate in the liquid phase, for example by a convective electrolyte flow or an optimization of the electrolyte flow field. Also, new requirements for high-performance ODCs can be formulated on the basis of

[11] Measurements performed by Institute of Chemical and Electrochemical Process Engineering, Technische Universität Clausthal, Germany and Analytical Chemistry — Center for Electrochemical Sciences, Ruhr University Bochum, Germany, published in [2].

[12] Measurements performed by Institute of Chemical and Electrochemical Process Engineering, Technische Universität Clausthal, Germany, published in [15].

the investigations. The structure of the ODC must not only support a large size of the gas-liquid interface, but also a location of the gas-liquid interface close to the liquid side of the ODC.

From a methodological point of view, the use of dynamic methods is indispensable, as this enables the processes to be separated from each other by their time constants and to determine their specific influence on electrode performance. Stationary methods are limited in this matter. For an investigation of the interaction between different processes and their effect on the electrode performance, a one-dimensional approach has been shown to be sufficient, since the dynamics of the different processes as well as their interaction can already be represented in this manner. However, this approach does not allow to represent inhomogeneities or parameter distributions; this is enabled through parallel interconnection of electrodes with different properties.
As was shown, for ODCs potential steps are particularly suitable for identification and analysis of processes with slower time constants as mass transport in the liquid phase. Impedance spectra are appropriate for a more detailed status analysis of the complex three-phase system due to their high information content with regard to electrolyte distribution or local reaction rates.

7.2 Open questions and perspective

In this dissertation, ODCs were analyzed by using macroscopic models. It was possible to identify performance drivers as well as limiting factors of ODCs. However, the results have also raised new questions, which can be answered in future studies by the utilization and further extension of the developed models:
For the ODC in the measurement setup with the forced convective electrolyte flow, a comparatively small thicknesses of the liquid diffusion layer was identified. A reason for this phenomenon may lay in additional mass transport effects, which are not yet covered by the model. There may be, for example, an additional natural convective flows perpendicular to the electrode due to density gradients as a result of concentration or temperature gradients. For the clarification of this aspect, e.g. CFD simulations would be a suitable diagnostic tool, whereby the macroscopic model can provide necessary data as ion concentration profiles for different current densities and states of the ODC.
Since the herein analyzed measurements were performed in temperature controlled setups, the power-related heating of the system was not considered. However, even at low current densities of $j = 3000\,\mathrm{A\,m^{-2}}$ at an overpotential of $\eta = 0.8\,\mathrm{V}$ (cf. fig. 5.2), the system is heated with a thermal power of $P = 2400\,\mathrm{W\,m^{-2}}$. Therefore, it is conceivable that the temperature is not ideally constant in terms of time and space. In the potential step measurements under certain conditions, a time-dependent behavior of the current was observed, which is opposite to the effect of water mass transport. Also divergences between the measurements and simulations at high current

densities were obtained. Both may be explained by temperature influences. To study the impact of this effect on electrode performance, the macroscopic model can be expanded to cover the temperature influences. For model validation combining electrochemical measurements with local temperature measurements would be suitable.

The macroscopic findings indicate that there are strong gradients in local performance within an ODC. To investigate the underlying mechanism and its relation to the ODC structure more deeply, simulations on the pore scale would be useful, taking effects as bottle necks and dead spaces within the pore network into account. For this purpose, the macroscopic ODC model could be transformed into a pore network model.

Bibliography

[1] M. Röhe, F. Kubannek, and U. Krewer, "Processes and their Limitations in Oxygen Depolarized Cathodes: A Dynamic Model - Based Analysis," *ChemSusChem*, vol. 11, pp. 2373–2385, 2019.

[2] M. Röhe, A. Botz, D. Franzen, F. Kubannek, B. Ellendorff, D. Öhl, W. Schuhmann, T. Turek, and U. Krewer, "The key role of water activity for the operating behavior and dynamics of oxygen depolarized cathodes," *ChemElectroChem*, vol. 6, no. 22, pp. 5671–5681, 2019.

[3] J. Kintrup, M. Millaruelo, V. Trieu, A. Bulan, and E. S. Mojica, "Gas Diffusion Electrodes for Efficient Manufacturing of Chlorine and Other Chemicals," *The Electrochemical Society Interface*, vol. 26, no. 2, pp. 73–76, 2017.

[4] C. H. Hamann and W. Vielstich, *Elektrochemie*. John Wiley & Sons Australia, Milton, Queensland, Limited, 2005.

[5] T. F. O'Brien, T. V. Bommaraju, and F. Hine, *Handbook of Chlor-Alkali Technology*. Boston, MA: Springer US, 2005.

[6] I. Moussallem, J. Jörissen, U. Kunz, S. Pinnow, and T. Turek, "Chlor-alkali electrolysis with oxygen depolarized cathodes: History, present status and future prospects," *Journal of Applied Electrochemistry*, vol. 38, no. 9, pp. 1177–1194, 2008.

[7] statista (A. Breitkopf), "Weltweiter Stromverbrauch in den Jahren 1980 bis 2017," 2021, https://de.statista.com/statistik/daten/studie/239764/umfrage/weltweiter-stromverbrauch/, date accessed: 25/01/2021. [Online]. Available: https://de.statista.com/statistik/daten/studie/239764/umfrage/weltweiter-stromverbrauch/

[8] N. Chavan, S. Pinnow, G. D. Polcyn, and T. Turek, "Non-isothermal model for an industrial chlor-alkali cell with oxygen-depolarized cathode," *Journal of Applied Electrochemistry*, vol. 45, no. 8, pp. 899–912, 2015.

[9] I. Moussallem, S. Pinnow, N. Wagner, and T. Turek, "Development of high-performance silver-based gas-diffusion electrodes for chlor-alkali electrolysis with oxygen depolarized

cathodes," *Chemical Engineering and Processing: Process Intensification*, vol. 52, pp. 125–131, 2012.

[10] Q. Mao and U. Krewer, "Total harmonic distortion analysis of oxygen reduction reaction in proton exchange membrane fuel cells," *Electrochimica Acta*, vol. 103, pp. 188–198, 2013.

[11] U. Krewer, M. Christov, T. Vidaković, and K. Sundmacher, "Impedance spectroscopic analysis of the electrochemical methanol oxidation kinetics," *Journal of Electroanalytical Chemistry*, vol. 589, no. 1, pp. 148–159, 2006.

[12] D. Schröder, V. Laue, and U. Krewer, "Numerical simulation of gas-diffusion-electrodes with moving gas-liquid interface: A study on pulse-current operation and electrode flooding," *Computers and Chemical Engineering*, vol. 84, pp. 217–225, 2016.

[13] U. Krewer, F. Röder, E. Harinath, R. D. Braatz, B. Bedürftig, and R. Findeisen, "Dynamic Models of Li-Ion Batteries for Diagnosis and Operation: A Review and Perspective," *Journal of The Electrochemical Society*, vol. 165, no. 16, pp. A3656–A3673, 2018.

[14] Q. Mao and U. Krewer, "Sensing methanol concentration in direct methanol fuel cell with total harmonic distortion: Theory and application," *Electrochimica Acta*, vol. 68, pp. 60–68, 2012.

[15] M. Röhe, D. Franzen, F. Kubannek, B. Ellendorff, T. Turek, and U. Krewer, "Revealing the degree and impact of inhomogeneous electrolyte distributions on silver based gas diffusion electrodes," *Electrochimica Acta*, vol. 389, p. 138693, 2021.

[16] P. Schmittinger, T. Florkiewicz, L. Curlin, B. Lüke, R. Scannell, T. Navin, Zelfel, E., and Bartsch, "Ceramics to Chlorohydrins," *Ullmann's Encyclopedia of Industrial Chemistry*, vol. A6, pp. 473 – 488, 1987.

[17] C. Butler, "Brine Electrolysis," 1954, uS Patent 2,681,884.

[18] J. Jörissen, T. Turek, and R. Weber, "Chlorherstellung mit Sauerstoffverzehrkathoden. Energieeinsparung bei der Elektrolyse," *Chemie in Unserer Zeit*, vol. 45, no. 3, pp. 172–183, 2011.

[19] J. Jung, S. Postels, and A. Bardow, "Cleaner chlorine production using oxygen depolarized cathodes? A life cycle assessment," *Journal of Cleaner Production*, vol. 80, pp. 46–56, 2014.

[20] R. Kuwertz, I. G. Martinez, T. Vidaković-Koch, K. Sundmacher, T. Turek, and U. Kunz, "Energy-efficient chlorine production by gas-phase HCl electrolysis with oxygen depolarized cathode," *Electrochemistry Communications*, vol. 34, pp. 320–322, 2013.

[21] M. Chatenet, L. Genies-Bultel, M. Aurousseau, R. Durand, and F. Andolfatto, "Oxygen reduction on silver catalysts in solutions containing various concentrations of sodium hydroxide - Comparison with platinum," *Journal of Applied Electrochemistry*, vol. 32, no. 10, pp. 1131–1140, 2002.

[22] D. Tromans, "Modeling Oxygen Solubility in Water and Electrolyte Solutions," *Industrial & Engineering Chemistry Research*, vol. 39, no. 3, pp. 805–812, 2000.

[23] D. Franzen, B. Ellendorff, M. C. Paulisch, A. Hilger, M. Osenberg, I. Manke, and T. Turek, "Influence of binder content in silver-based gas diffusion electrodes on pore system and electrochemical performance," *Journal of Applied Electrochemistry*, vol. 49, no. 7, pp. 705–713, 2019.

[24] P. K. Adanuvor, "Oxygen Reduction on Silver in 6.5M Caustic Soda Solution," *Journal of The Electrochemical Society*, vol. 135, no. 10, p. 2509, 1988.

[25] B. B. Blizanac, P. N. Ross, and N. M. Marković, "Oxygen reduction on silver low-index single-crystal surfaces in alkaline solution: Rotating ring disk$_{Ag(hkl)}$ studies," *Journal of Physical Chemistry*, vol. 110, no. 10, pp. 4735–4741, 2006.

[26] B. B. Blizanac, P. N. Ross, and N. M. Markovic, "Oxygen electroreduction on Ag(1 1 1): The pH effect," *Electrochimica Acta*, vol. 52, no. 6, pp. 2264–2271, 2007.

[27] C. Zhang, F. R. F. Fan, and A. J. Bard, "Electrochemistry of oxygen in concentrated NaOH solutions: Solubility, diffusion coefficients, and superoxide formation," *Journal of the American Chemical Society*, vol. 131, no. 1, pp. 177–181, 2009.

[28] A. Ignaczak, R. Nazmutdinov, A. Goduljan, L. M. de Campos Pinto, F. Juarez, P. Quaino, E. Santos, and W. Schmickler, "A scenario for oxygen reduction in alkaline media," *Nano Energy*, vol. 29, pp. 362–368, 2016.

[29] S. Kandaswamy, A. Sorrentino, S. Borate, L. A. Živković, M. Petkovska, and T. Vidaković-Koch, "Oxygen reduction reaction on silver electrodes under strong alkaline conditions," *Electrochimica Acta*, vol. 320, p. 134517, 2019.

[30] X. Ge, A. Sumboja, D. Wuu, T. An, B. Li, F. W. T. Goh, T. S. A. Hor, Y. Zong, and Z. Liu, "Oxygen Reduction in Alkaline Media: From Mechanisms to Recent Advances of Catalysts," *ACS Catalysis*, vol. 5, no. 8, pp. 4643–4667, 2015.

[31] E. L. Littauer, "Catalytic Decomposition of Hydrogen Peroxide in Alkaline Solution," *Journal of The Electrochemical Society*, vol. 126, no. 11, p. 1924, 1979.

[32] S. Pinnow, N. Chavan, and T. Turek, "Thin-film flooded agglomerate model for silver-based oxygen depolarized cathodes," *Journal of Applied Electrochemistry*, vol. 41, no. 9, pp. 1053–1064, 2011.

[33] N. Ramaswamy and S. Mukerjee, "Fundamental mechanistic understanding of electrocatalysis of oxygen reduction on Pt and non-Pt surfaces: Acid versus alkaline media," *Advances in Physical Chemistry*, vol. 2012, pp. 1–17, 2012.

[34] A. Botz, J. Clausmeyer, D. Öhl, T. Tarnev, D. Franzen, T. Turek, and W. Schuhmann, "Local Activities of Hydroxide and Water Determine the Operation of Silver-Based Oxygen Depolarized Cathodes," *Angewandte Chemie International Edition*, vol. 57, pp. 12 285–12 289, 2018.

[35] U. Krewer, H.-K. Yoon, and H.-T. Kim, "Basic model for membrane electrode assembly design for direct methanol fuel cells," *Journal of Power Sources*, vol. 175, no. 2, pp. 760–772, 2008.

[36] S. Chevalier, B. Auvity, J. C. Olivier, C. Josset, D. Trichet, and M. Machmoum, "Detection of cells state-of-health in PEM fuel cell stack using EIS measurements coupled with multiphysics modeling," *Fuel Cells*, vol. 14, no. 3, pp. 416–429, 2014.

[37] U. Krewer, A. Kamat, and K. Sundmacher, "Understanding the dynamic behaviour of direct methanol fuel cells: Response to step changes in cell current," *Journal of Electroanalytical Chemistry*, vol. 609, no. 2, pp. 105 – 119, 2007.

[38] C. Y. Yuh and J. R. Selman, "Polarization of the Molten Carbonate Fuel Cell Anode and Cathode," *Journal of The Electrochemical Society*, vol. 131, no. 9, pp. 2062–2069, 1984.

[39] S. Pinnow, "Die Modellierung von Sauerstoffverzehr- Kathoden für Chloralkali-Elektrolyse," Dissertation, Technische Universität Clausthal, 2013.

[40] R. C. Burshtein, V. S. Markin, A. G. Pshenichnikov, V. A. Chismadgev, and Y. G. Chirkov, "The relationship between structure and electrochemical properties of porous gas electrodes," *Electrochimica Acta*, vol. 9, no. 6, pp. 773–787, 1964.

[41] J. Giner and C. Hunter, "The Mechanism of Operation of the Teflon-Bonded Gas Diffusion Electrode: A Mathematical Model," *Journal of The Electrochemical Society*, vol. 116, no. 8, p. 1124, 1969.

[42] M. B. Cutlip, "An approximate model for mass transfer with reaction in porous gas diffusion electrodes," *Electrochimica Acta*, vol. 20, pp. 767–773, 1975.

[43] L. G. Austin, M. Ariet, R. D. Walker, G. B. Wood, and R. H. Comyn, "Simple-pore and thin-film models of porous gas diffusion electrodes," *Industrial and Engineering Chemistry Fundamentals*, vol. 4, no. 3, pp. 321–327, 1965.

[44] J. T. Cobb and L. F. Albright, "The Effect of Preoxidation and Meniscus shape on the Hydrogen-Platinum Anode of Molten-Carbonate Fuel Cell," *Journal of the Electrochemical Society*, vol. 115, no. 1, pp. 2 – 6, 1968.

[45] F. G. Will and D. J. Bendaniel, "Significance of Electrolyte Films for Performance of Porous Hydrogen Electrodes," *General Electric Research and Development Center*, vol. 116, pp. 933–937, 1969.

[46] F. Kubannek, T. Turek, and U. Krewer, "Modeling Oxygen Gas Diffusion Electrodes for Various Technical Applications," *Chemie-Ingenieur-Technik*, vol. 91, no. 6, pp. 720–733, 2019.

[47] X.-L. Wang and S. Koda, "Scale-up and Modeling of Oxygen Diffusion Electrodes for Chlorine-Alkali Electrolysis II. Effects of the Structural Parameters on the Electrode Performance Based on the Thin-Film and Flooded-Agglomerate Model," *DENKI KAGAKU*, vol. 65, no. 12, pp. 1014 –1025, 1997.

[48] M. Sudoh, K. Arai, Y. Izawa, T. Suzuki, M. Uno, M. Tanaka, K. Hirao, and Y. Nishiki, "Evaluation of Ag-based gas-diffusion electrode for two-compartment cell used in novel chlor-alkali membrane process," *Electrochimica Acta*, vol. 56, no. 28, pp. 10 575–10 581, 2011.

[49] J. Lindberg, B. Endródi, G. Åvall, P. Johansson, A. Cornell, and G. Lindbergh, "Li Salt Anion Effect on O_2-Solubility in an Li-O_2 Battery," *Journal of Physical Chemistry C*, vol. 122, no. 4, pp. 1913–1920, 2018.

[50] S. L. Clegg and P. Brimblecombe, "The solubility and activity coefficient of oxygen in salt solutions and brines," vol. 54, no. 4, pp. 3315–3328, 1990.

[51] H. G. Hirschberg, *Handbuch Verfahrenstechnik und Anlagenbau.* Springer-Verlag Berlin Heidelberg New York, 1999.

[52] J. Balej, "Water vapour partial pressures and water activities in potassium and sodium hydroxide solutions over wide concentration and temperature ranges," *International Journal of Hydrogen Energy*, vol. 10, no. 4, pp. 233–243, 1985.

[53] A. J. Bard and L. R. Faulkner, *Electrochemical Methods: Fundamentals and Applications*, 2nd ed. John Wiley & Sons, INC., New York, 2000.

[54] B. Kögl and F. Moser, *Grundlagen der Verfahrenstechnik.* Vienna: Springer Vienna, 1981.

[55] J. Newman and K. E. Thomas-Alyea, *Electrochemical Systems*, 3rd ed. New York: John Wiley & Sons, 2012.

[56] R. Krishna, "Problems and pitfalls in the use of the fick formulation for intraparticle diffusion," *Chemical Engineering Science*, vol. 48, no. 5, pp. 845–861, 1993.

[57] B. E. Poling, J. M. Prausnitz, and J. P. O'Connell, *The Properties of Gases and Liquids*, 5th ed., ser. McGraw Hill professional. McGraw-Hill Education, 2000.

[58] R. Krishna and J. Wesselingh, "The Maxwell-Stefan approach to mass transfer," *Chemical Engineering Science*, vol. 52, no. 6, pp. 861–911, 1997.

[59] R. Taylor and R. Krishna, *Multicomponent Mass Transfer*, ser. Wiley Series in Chemical Engineering. John Wiley & Sons, INC., New York, 1993.

[60] J. Newman, D. Bennion, and C. W. Tobias, "Mass Transfer in Concentrated Binary Electrolytes," *Berichte der Bunsengesellschaft für physikalische Chemie*, vol. 69, pp. 608–612, 1965.

[61] D. S. Christen, *Praxiswissen der chemischen Verfahrenstechnik.* Berlin, Heidelberg: Springer, 2010.

[62] M. E. Orazem and B. Tribollet, *Electrochemical Impedance Spectroscopy.* John Wiley & Sons, Inc., Hoboken, New Jersey, 2008.

[63] D. Öhl, D. Franzen, M. Paulisch, S. Dieckhöfer, S. Barwe, C. Andronescu, I. Manke, T. Turek, and W. Schuhmann, "Catalytic Reactivation of Industrial Oxygen Depolarized Cathodes by in situ Generation of Atomic Hydrogen," *ChemSusChem*, vol. 12, no. 12, pp. 2732–2739, 2019.

[64] T. Reshetenko, V. Laue, V. Krewer, and K. Artyushkova, "Poisoning effects of sulfur dioxide in an air stream on spatial proton exchange membrane fuel cell performance," *Journal of Power Sources*, vol. 438, p. 226949, 2019.

[65] I. D. Raistrick, D. R. Franceschetti, and J. R. Macdonald, *Theory.* John Wiley & Sons, INC., Hoboken, New Jersey, 2005, ch. 2, pp. 27–128.

[66] J. Balej, "Activity coefficients of aqueous solutions of NaOH and KOH in wide concentration and temperature ranges," *Collection of Czechoslovak Chemical Communications*, vol. 61, pp. 1549–1562, 1996.

[67] J. van Brakel and P. M. Heertjes, "Analysis of diffusion in macroporous media in terms of a porosity, a tortuosity and a constrictivity factor," *International Journal of Heat and Mass Transfer*, vol. 17, no. 9, pp. 1093–1103, 1974.

[68] J. Oisson, I. A. Jernqvist, and G. I. Aly, "Thermophysical Properties of Aqueous NaOH-H_2O Solutions at High Concentrations," *International Journal of Thermophysics*, vol. 18, no. 3, 1997.

[69] M. Neumann, M. Osenberg, A. Hilger, D. Franzen, T. Turek, I. Manke, and V. Schmidt, "On a pluri-Gaussian model for three-phase microstructures, with applications to 3D image data of gas-diffusion electrodes," *Computational Materials Science*, vol. 156, no. September 2018, pp. 325–331, 2019.

[70] J. Clausmeyer, A. Botz, D. Öhl, and W. Schuhmann, "The oxygen reduction reaction at the three-phase boundary: nanoelectrodes modified with Ag nanoclusters," *Faraday Discussions*, vol. 193, pp. 241–250, 2016.

[71] S. G. Bratsch, "Standard Electrode Potentials and Temperature Coefficients in Water at 298.15 K," *Journal of Physical and Chemical Reference Data*, vol. 18, no. 1, pp. 1–21, 1989.

[72] G. D. Polcyn, *Charakterisierung und Modellierung der Sauerstoffreduktion an Sauerstoff-Verzehr-Kathoden für die Chlor-Alkali-Elektrolyse anhand von Abschaltmessungen.* Verlag Dr. Hut, München, 2019, technische Universität Dortmund, *Dissertation*.

[73] B. E. Poling, J. Prausnitz, and J. O'Connell, *The Properties of Gases and Liquids*, ser. McGraw Hill professional. McGraw-Hill Education, 2000.

[74] M. Chatenet, M. Aurousseau, and R. Durand, "Comparative methods for gas diffusivity and solubility determination in extreme media: Application to molecular oxygen in an industrial chlorine-soda electrolyte," *Industrial and Engineering Chemistry Research*, vol. 39, no. 8, pp. 3083–3089, 2000.

[75] G. Åkerlöf and G. Kegeles, "Thermodynamics of Concentrated Aqueous Solutions of Sodium Hydroxide," *Journal of the American Chemical Society*, vol. 62, no. 3, pp. 620–640, 1940.

[76] T. Danner, B. Horstmann, D. Wittmaier, N. Wagner, and W. G. Bessler, "Reaction and transport in Ag/Ag2O gas diffusion electrodes of aqueous Li-O2 batteries: Experiments and modeling," *Journal of Power Sources*, vol. 264, pp. 320–332, 2014.

[77] M. J. Blandamer, J. B. F. N. Engberts, P. T. Gleeson, and J. C. R. Reis, "Activity of water in aqueous systems; a frequently neglected property," *Chemical Society Reviews*, vol. 34, no. 5, pp. 440–458, 2005.

[78] A. Botti, F. Bruni, S. Imberti, M. A. Ricci, and A. K. Soper, "Ions in water: The microscopic structure of concentrated NaOH solutions," *Journal of Chemical Physics*, vol. 120, no. 21, pp. 10 154–10 162, 2004.

[79] A. Botti, F. Bruni, S. Imberti, M. A. Ricci, A. K. Soper, "Solvation of hydroxyl ions in water," *Journal of Chemical Physics*, vol. 119, no. 10, pp. 5001–5004, 2003.

[80] M. C. Paulisch, M. Gebhard, D. Franzen, A. Hilger, M. Osenberg, N. Kardjilov, B. Ellendorff, T. Turek, C. Roth, and I. Manke, "Operando laboratory x-ray imaging of silver-based gas diffusion electrodes during oxygen reduction reaction in highly alkaline media," *Materials*, vol. 12, no. 17, 2019.

[81] P. Kunz, M. Hopp-Hirschler, and U. Nieken, "Simulation of Electrolyte Imbibition in Gas Diffusion Electrodes," *Chemie Ingenieur Technik*, vol. 91, no. 6, pp. 883–888, 2019.

[82] S. Tsushima, K. Teranishi, K. Nishida, and S. Hirai, "Water content distribution in a polymer electrolyte membrane for advanced fuel cell system with liquid water supply," *Magnetic Resonance Imaging*, vol. 23, no. 2 SPEC. ISS., pp. 255–258, 2005.

[83] P. Krüger, H. Markötter, J. Haußmann, M. Klages, T. Arlt, J. Banhart, C. Hartnig, I. Manke, and J. Scholta, "Synchrotron X-ray tomography for investigations of water distribution in polymer electrolyte membrane fuel cells," *Journal of Power Sources*, vol. 196, no. 12, pp. 5250–5255, 2011.

[84] B. Tavakoli and R. Roshandel, "The effect of fuel cell operational conditions on the water content distribution in the polymer electrolyte membrane," *Renewable Energy*, vol. 36, no. 12, pp. 3319–3331, 2011.

[85] M. D. Levi, C. Wang, and D. Aurbach, "Two parallel diffusion paths model for interpretation of PITT and EIS responses from non-uniform intercalation electrodes," *Journal of Electroanalytical Chemistry*, vol. 561, pp. 1–11, 2004.

[86] K. Jüttner and W. Lorenz, "Electrochemical Impedance Spectroscopy (EIS)of Corrosion Processes on Inhomogeneous Surfaces," *Materials Science Forum*, vol. 44-45, pp. 191–204, 1991.

[87] U. Krewer, A. Kamat, and K. Sundmacher, "Understanding the dynamic behaviour of direct methanol fuel cells: Response to step changes in cell current," *Journal of Electroanalytical Chemistry*, vol. 609, no. 2, pp. 105–119, 2007.

[88] X. Zhong, M. Schulz, C. H. Wu, M. Rabe, A. Erbe, and M. Rohwerder, "Limiting Current Density of Oxygen Reduction under Ultrathin Electrolyte Layers: From the Micrometer Range to Monolayers," *ChemElectroChem*, vol. 8, no. 4, pp. 712–718, 2021.

[89] T. Muzaffar, T. Kadyk, and M. Eikerling, "Tipping water balance and the Pt loading effect in polymer electrolyte fuel cells: A model-based analysis," *Sustainable Energy and Fuels*, vol. 2, no. 6, pp. 1189–1196, 2018.

[90] Z. Lukács, "The numerical evaluation of the distortion of EIS data due to the distribution of parameters," *Journal of Electroanalytical Chemistry*, vol. 432, no. 1-2, pp. 79–83, 1997.

[91] A. Schröder, K. Wippermann, W. Lehnert, D. Stolten, T. Sanders, T. Baumhöfer, N. Kardjilov, A. Hilger, J. Banhart, and I. Manke, "The influence of gas diffusion layer wettability on direct methanol fuel cell performance: A combined local current distribution and high resolution neutron radiography study," *Journal of Power Sources*, vol. 195, no. 15, pp. 4765–4771, 2010.

[92] A. A. Kulikovsky, J. Divisek, and A. A. Kornyshev, "Modeling the Cathode Compartment of Polymer Electrolyte Fuel Cells: Dead and Active Reaction Zones," *Journal of The Electrochemical Society*, vol. 146, no. 11, pp. 3981–3991, 1999.

[93] R. B. Bird and D. J. Klingenberg, "Multicomponent diffusion – A brief review," *Advances in Water Resources*, vol. 62, pp. 238–242, 2013.

[94] D. B. Sepa, M. V. Vojnovic, L. M. Vracar, and A. Damjanovic, "Apparent enthalpies of activation of electrodic oxygen reduction at platinum in different current density regions-II. Alkaline solution," *Electrochimica Acta*, vol. 31, no. 1, pp. 97–101, 1986.

[95] J. N. Soderberg, A. C. Co, A. H. Sirk, and V. I. Birss, "Impact of porous electrode properties on the electrochemical transfer coefficient," *Journal of Physical Chemistry B*, vol. 110, no. 21, pp. 10 401–10 410, 2006.

[96] H. Lindner, *Physik für Ingenieure*, 18th ed. Hanser Fachbuchverlag, München, 2010.

[97] J.-L. Burgot, *The Notion of Activity in Chemistry*. Cham: Springer International Publishing, 2017.

[98] N. Nioradze, R. Chen, J. Kim, M. Shen, P. Santhosh, and S. Amemiya, "Origins of nanoscale damage to glass-sealed platinum electrodes with submicrometer and nanometer size," *Analytical Chemistry*, vol. 85, no. 13, pp. 6198–6202, 2013.

[99] B. Ballesteros Katemann and W. Schuhmann, "Fabrication and Characterization of Needle-Type Pt-Disk Nanoelectrodes," *Electroanalysis*, vol. 14, no. 1, pp. 22–28, 2002.

[100] R. Brunner, A. Bietsch, O. Hollricher, and O. Marti, "Distance control in near-field optical microscopy with piezoelectrical shear-force detection suitable for imaging in liquids," *Review of Scientific Instruments*, vol. 68, no. 4, pp. 1769–1772, 1997.

[101] M. Ludwig, C. Kranz, W. Schuhmann, and H. E. Gaub, "Topography feedback mechanism for the scanning electrochemical microscope based on hydrodynamic forces between tip and sample," *Review of Scientific Instruments*, vol. 66, no. 4, pp. 2857–2860, 1995.

[102] M. Nebel, K. Eckhard, T. Erichsen, A. Schulte, and W. Schuhmann, "4D shearforce-based constant-distance mode scanning electrochemical microscopy," *Analytical Chemistry*, vol. 82, no. 18, pp. 7842–7848, 2010.

List of Figures

1.1 Schematic view on processes in oxygen depolarized cathodes. 2
2.1 Schematic illustration of the chlor-alkali electrolysis membrane processes, a) without and b) with ODC, based on [3, 6]. 8
3.1 Schematic illustration of different model approaches for three-phase systems within a porous GDE: a) single-pore model (flooded), based on [38]; b) meniscus-model, based on [38]; c) thin-film model, based on [38]; d) model with micro and macro pores, based on [40]; e), flooded agglomerates model, based on [41] f); thin-film flooded agglomerate (TFFA) model, based on [42]. 14
3.2 a) NaOH concentration dependent inverse Henry constant at 80 °C, calculated based on [22]; b) NaOH concentration dependent normalized water vapor pressure at 80 °C, calculated based on [51]. 17
3.3 Dependency of current density on overpotential close to the OCP, calculated by Bulter-Volmer and Tafel equation. 19
3.4 Schematic illustration of Maxwell-Stefan diffusion for the binary case. Partly based on [56]. 22
3.5 Schematic illustration of a) bilateral equimolar diffusion (based on [54]); b) bilateral non-equimolar diffusion; c) unilateral diffusion. 25
3.6 Time constants of processes in ODCs and suitable methods to detect them, Δz is the characteristic diffusion length, partly based on [13, 35]. 28
3.7 Schematic illustration of changes in states during a potential step. 30
3.8 Schematic illustration of electrochemical impedance spectroscopy analysis and the related dynamic state changes. 32
4.1 Schematic view of the model including reaction and mass transport. 38
4.2 Mass transport processes at the gas-liquid interface. 38
4.3 Illustration of model on water evaporation and oxygen dissolution immediately at the gas-liquid interface. While the oxygen is solved from the gas phase into the liquid electrolyte (top), the water evaporates from the liquid electrolyte into the gas phase (bottom). 41
4.4 Simulated polarization curve in comparison to literature data [32]. 46
4.5 Dependence of simulated a) oxygen and b) water concentrations profiles within the liquid phase on current density. 49

4.6 Fraction of mass flow driven by diffusion in comparison to the total mass flow in the flooded agglomerats as a function the current density at various locations. 50
4.7 Electrode potential and current dependent on the time for a potential step of -0.55 to -0.65 V vs. OCP at $t = 10$ s and back at $t = 20$ s. 50
4.8 a) Concentration responses of oxygen and water at the thin-film-liquid interface for a potential step of -0.55 to -0.65 V vs. OCP at $t = 10$ s and back at $t = 20$ s and b) response of oxygen partial pressure at various locations. 52
4.9 Relative sensitivity s_ς of the electrode potential to parameter variation of $\pm$ 5% for three different constant current densities. When increasing c_{NaOH}, the current density of $j = 8\,kA\,m^{-2}$ could not be reached, thus only results for negative variation are shown. 54
5.1 Schematic view of the used measuring cells with a) convective electrolyte flow and b) stagnant electrolyte. 60
5.2 Simulated polarization curves in comparison to experimental polarization curves for convective electrolyte flow and stagnant electrolyte cell. 65
5.3 Dependence of simulated oxygen concentrations profiles within the liquid phase with respect to the current density for a) convective electrolyte flow and b) stagnant electrolyte. 66
5.4 Simulated and experimental NaOH concentration in the liquid diffusion layer 1 µm above the ODC surface for cell with stagnant electrolyte. For comparison, NaOH concentration in the cell with the convective electrolyte flow is also given at the same location. 67
5.5 Simulated and experimental time-dependent current density response for a potential step from 0.8 to 0.7 V vs. RHE at $t = 30$ s and a) back at $t = 60$ s for convective electrolyte flow cell and b) back at $t = 90$ s for stagnant electrolyte cell. 68
5.6 Simulated NaOH concentration and water activity within the liquid phase dependent on the time for a potential step from 0.8 to 0.7 V vs. RHE at $t = 30$ s and a) back at $t = 60$ s for convective electrolyte flow cell and b) back at $t = 90$ s for stagnant electrolyte cell. 70
5.7 Henry constant at the gas-liquid interface (left) at steady state current at 0.7 V vs. RHE and its change at selected time points after a potential step to 0.8 V vs. RHE. Corresponding gradient of water activity within the ODC across the flooded agglomerates (middle) and in the liquid diffusion layer (right). Cell with no electrolyte convection. 71

5.8 Simulated oxygen concentration within the liquid phase dependent on the time for a potential step from 0.8 to 0.7 V vs. RHE at $t = 30$ s and a) back at $t = 60$ s for convective electrolyte flow cell and b) back at $t = 90$ s for stagnant electrolyte cell (fa: flooded agglomerates). 72
6.1 Experimental [23] and simulated polarization curves of ODCs with different Ag-content. 78
6.2 a) Simulated and experimental EIS at $j = 2000\,\mathrm{A\,m^{-2}}$ of ODCs with different Ag-content, b) calculated charge transfer resistance and c) calculated time constants from experimental and simulated EIS. 79
6.3 a) Schematic view of the model of an inhomogeneous ODC comprised of multiple homogeneous sub-domains; b) schematic view of the domain sub-model including reaction and mass transport; c) zoom at the gas-liquid interface. 83
6.4 Simulated current density dependent on the potential in each domain k and for total ODC, for ODCs with a) 92 wt−% Ag and b) 98 wt−% Ag. 93
6.5 Identified properties of ODCs with different Ag-content: a) Electrolyte distribution dependent on the dimensionless location within the gas diffusion electrode (calculated by eq. (6.40)); b) average area of gas-liquid interface per electrode volume dependent on the dimensionless location within the gas diffusion electrode (calculated by eq. (6.41)); c) Schematic illustration of the electrolyte intrusion and gas-liquid interface distribution for reference. 95
6.6 Bode plots for EIS at $j = 2000\,\mathrm{A\,m^{-2}}$ for ODC with 98 wt−% Ag for full electrode and single electrode domains. The three scenarios: a), b) inhomogeneous ODC; c), d) homogeneous ODC; e), f) semi-homogeneous ODC. 98
6.7 Experimental and different simulated scenarios: a) EIS at $j = 2000\,\mathrm{A\,m^{-2}}$ for ODC with 98 wt−% Ag; b) polarization curve. 100
6.8 Different simulated scenarios: a) Time-dependent current density response for a potential step from 0.8 to 0.7 V vs. RHE at $t = 10$ s and back at $t = 20$ s; b) and c): Zooms of a). 101
6.9 Time-dependent capacitiv current density response for a potential step from 0.8 to 0.7 V vs. RHE at $t = 10$ s for full electrode and single electrode domains of the inhomogeneous ODC. 102
B.1 Total, water, and sodium hydroxide concentration of liquid electrolyte, calculated based on Hirschberg [51] and implemented in the model, depended on the modeled water concentration in the range of the simulation at $T = 353.15$ K. 133

B.2 Inverse Henry constant over water concentration at $T = 353.15\,\mathrm{K}$, hydroxide concentration according to figure B.1. 134
B.3 Vapor pressure of water as a function of water concentration at $T = 353.15\,\mathrm{K}$, hydroxide concentration according to figure B.1. 134
B.4 Activity of water as a function of water concentration at $T = 353.15\,\mathrm{K}$, hydroxide concentration according to figure B.1. 135
B.5 Density of electrolyte as a function of water concentration at $T = 353.15\,\mathrm{K}$, hydroxide concentration according to figure B.1. 135
D.1 Simulated and experimental EIS at different current densities of ODC with 98 wt−% Ag. 143

List of Tables

3.1 List of properties varying in the different model expansion stages. 34
4.1 List of constant parameter and operation conditions. 47
4.2 Electrolyte concentration depended variables. 47
4.3 Identified parameters. 48
5.1 Electrode characteristics. 59
5.2 Model equations for expanded liquid mass transport model. (Model schema for reference: fig: 4.1.) 61
5.3 List of constant parameters and operating conditions. 64
5.4 Electrolyte concentration dependent variables. 64
5.5 Parameters identified from electrochemical experiments. 65
6.1 Model equations for domain-sub and inhomogeneous model. (Model schema for reference: fig. 6.3.) 84
6.2 Geometrical dimensions of the three-phase area of the different domains. 88
6.3 Parameters identified from (electrochemical) experiments and ODC model for electrodes with different Ag content as well as literature data. 90
6.4 Distributed parameters identified from electrochemical experiments and ODC model for electrodes with different Ag content. 90
6.5 List of constant parameters and operating conditions. 91
6.6 Electrolyte concentration-dependent variables. 92
D.1 Parameters identified from electrochemical experiments and ODC model for ODC with 98 wt−% Ag with hypothetical scenarios. 144

List of Symbols and Abbreviations

Latin letters

A	Area	m^2
a_i	Activity of species i	–
b_i	Molality of species i	$mol\,kg^{-1}$
C_{dl}	Double layer capacitance	$F\,m^{-2}$
c_i	Concentration of species i	$mol\,m^{-3}$
D_{ij}	Binary diffusion coefficient for species i and j (unspecified)	$m^2\,s^{-1}$
d_i	Coefficient for the forces acting on the molecules of species i	m^{-1}
$\mathscr{D}_{ij}$	Binary Fick's diffusion coefficient for species i and j	$m^2\,s^{-1}$
Đ$_{ij}$	Binary Maxwell-Stefan diffusion coefficient for species i and j	$m^2\,s^{-1}$
E	Potential	V
E_0	Open circuit potential	V
E_{00}	Standard electrode potential	V
F	Faraday constant (= 96485.34)	$C\,mol^{-1}$
H	(Inverse) Henry constant	$Pa\,m^3\,mol^{-1}$
j	Current density	$A\,m^{-2}$
k_0	Reaction rate constant	$mol\,s^{-1}$
$\dot{N}_i$	Mass flow density of species i	$mol\,s^{-1}\,m^{-2}$
p	Pressure	Pa
p_i	Partial pressure of species i	Pa
p_v	Proportion of the flooded pore volume	–
$\Delta p_{m,i,j}$	Exchanged momentum between species 1 and 2	$kg\,m\,s^{-1}$
R	Universal gas constant (= 8.3145)	$J\,mol^{-1}\,K^{-1}$
$R_{specific}$	Specific resistance	$\Omega\,m^{-2}$
R_{ct}	Charge transfer resistance	Ω

r	Reaction rate	$\mathrm{mol\,s^{-1}\,m^{-2}}$
S_{cat}	Specific surface area of the Ag catalyst per electrode volume	$\mathrm{m^2\,m^{-3}}$
$S_{\mathrm{domain,k}}$	Specific surface area of domain k per geometrical electrode area of the domain k	$\mathrm{m^2\,m^{-2}}$
$S_{\mathrm{phase,interface}}$	Specific surface area of the interface per geometrical electrode area	$\mathrm{m^2\,m^{-2}}$
T	Temperature	K
t	Time	s
v	Velocity	$\mathrm{m\,s^{-1}}$
x_i	Mole fraction of species i	–
z	Length	m
Z	Impedance	Ω

Greek letters

α	Charge transfer coefficient	–
ε	Porosity	–
γ_i	Activity coefficient of species i	–
$\gamma_{\pm}$	Mean activity coefficient of NaOH	–
η	Overpotential	V
Θ	Loading	$\mathrm{mg\,cm^{-2}}$
ν_i^{ORR}	Stoichiometric coefficient of species i in ORR	–
μ_i	Chemical potential of species i	$\mathrm{J\,mol^{-1}}$
τ	Time constant	s
τ	Tortuosity	–

Sub- and superscripts

3PA	Three-phase area
cat	Related to the catalyst surface
con	Convective electrolyte flow
c	Capacitive
DL	Double layer

d	Distance from the liquid side of the ODC
eff	Effective
ext	External
F	Faradaic
fa	Flooded agglomerates
gas	Gaseous / gas phase
geo	Geometrical / related to the geometric ODC surface
int	Internal
I	Imaginary
k	Index of domain
l	Thickness of ODC
liq	Liquid / liquid phase
ldl	Liquid diffusion layer
R	Real
stag	Stagnant electrolyte
t	Total
tf	Thin-film
$\circ$	Standard state
$*$	At gas-liquid interface

Abbreviations

3PA	Three phase area
BVE	Butler-Volmer-Equation
CFD	Computational fluid dynamics
DMFC	Direct methanol fuel cells
EIS	Electrochemical impedance spectroscopy
FA	Flooded agglomerates
GDE	Gas diffusion electrode
G	Gas phase
HER	Hydrogen evolution reaction
LSV	Linear sweep voltammetry
Liq	Liquid

LDL	Liquid diffusion layer
OCP	Open circuit potential
OCV	Open circuit voltage
ODC	Oxygen depolarized cathodes
ORR	Oxygen reduction reaction
PTFE	Polytetrafluoroethylene
PEMFC	Proton-exchange membrane fuel cell
RDS	Rate determing step
RHE	Reversible hydrogen electrode
SECM	Scanning electrochemical microscopy
SHE	Standard hydrogen electrode
TF	Thin-film
TFFA	Thin-film flooded agglomerate

A Maxwell-Stefan Diffusion

The stationary Maxwell-Stefan diffusion

The following exposition is mainly based on *Multicomponent mass transfer* by Taylor and Krischna [59] and *Mass transfer in concentrated binary electrolytes* by Newman et al. [60]. The execution is constrained on the one-dimensional case.

To understand the Maxwell-Stefan relations and to adapt this approach to the specific system of the ODCs, first the mechanism of molecular collisions have to be understood. The principle of molecular interaction is initially presented simplified for a binary system and generalized afterwards. The molecules of the species 1 (e.g. water) and 2 (e.g. hydroxide ions) move with an average velocity of v_1 and v_2. Introducing further the mass of a single molecule as m_1 and m_2, the momentum of the average molecule of both species can be written as $v_1 \cdot m_1$ and $v_2 \cdot m_2$. In case of a bilateral collision of molecules of both species, the velocity of both molecules changes, where v_1' and v_2' are the velocities after the collision. Thereby the total momentum $v_1 \cdot m_1 + v_2 \cdot m_2$ is conserved [59]:

$$m_1 \cdot (v_1 - v_1') + m_2 \cdot (v_2 - v_2') = 0 \tag{A.1}$$

From eq. A.1, the momentum transferred form the molecule of species 1 to the molecule of species 2 (and the other way round) can be calculated by [59]:

$$m_1 \cdot (v_1 - v_1') = m_2 \cdot (v_2 - v_2') \tag{A.2}$$

For determining the velocity of the molecules after collision, an ineleastic collision is assumed, where the molecule 1 and 2 stick together and move with the velocity $v_1' = v_2' = v'$, where v' is calculated by [59]:

$$v' = \frac{m_1 \cdot v_1 + m_2 \cdot v_2}{m_1 + m_2} \tag{A.3}$$

From eq. (A.3) the calculation of the exchanged momentum of the collision of two single molecules eq. (A.2) can be enlarged to eq. (A.4). Now it is shown in eq. (A.4), that the exchanged momentum is proportional to the difference of the velocity of both species $(v_1 - v_2)$ [59].

$$m_1 \cdot (v_1 - v_1') = m_2 \cdot (v_2 - v_2') = \frac{m_1 \cdot m_2}{m_1 + m_2}(v_1 - v_2) \tag{A.4}$$

Three important points can be extracted from this explanation so far [59]:

- When two molecules of different species collide, they exchange momentum.
- Although the momentum is conserved, the total kinetic energy is lowered and mainly converted to heat [96]:

  $$\frac{1}{2}(m_1 \cdot v_1^2 + m_2 \cdot v_2^2) > \frac{1}{2}(m_1 + m_2) \cdot v'^{\,2} \tag{A.5}$$

  It should be noted, however, that the kinetic energy of one of both species could increase. This implies that both positive and negative accelerations of molecules can be achieved through interaction.
- When two molecules of the same species meet, there is no loss of momentum for this species [59]. Accordingly, the momentum exchange between the molecules of a species does not have to be taken into account.

To transfer these considerations from single molecules to a macroscopic level, the frequency of collisions has to be respected: The number of collisions of molecules of the species 1 and 2 per unit and time is proportional to the product of their mole fractions $x_1 \cdot x_2$, with $x_1 = c_1/c_{\text{total}}$ and $x_2 = c_2/c_{\text{total}}$, where c is the concentration and $c_{\text{total}} = c_1 + c_2$ is to total concentration [59].
The two approaches can be combined: On a macroscopic level, the exchanged momentum $\Delta p_{\text{m},1,2}$ between species 1 and species 2 is proportional to the exchanged momentum at the collision of two single molecules and the frequency of collision [59]:

$$\Delta p_{\text{m},1,2} \propto ((x_1 \cdot x_2), (v_1 - v_2)) \tag{A.6}$$

Introducing now the Maxwell-Stefan diffusion coefficient $Đ_{1,2}$ as the inverse drag coefficient, the quantity $d_{1,2}$ is obtained [59]. Where $d_{1,2} \cdot \text{R}T \cdot c_{\text{total}}$ is the force per volume which species 2 is applying on species 1 resulting from the relative motion of species 1 and species 2 [60]:

$$d_{1,2} = -\frac{x_1 x_2 \cdot (v_1 - v_2)}{Đ_{1,2}} \tag{A.7}$$

A short excursion to Newton's second law shows that the force species 2 is applying on species 1, is the same force as species 1 is applying on species 2 [59].

$$d_{1,2} = -d_{2,1} \tag{A.8}$$

where

$$Đ_{1,2} = Đ_{2,1} \tag{A.9}$$

To generalize equation (A.7) from a binary to a multi-component mixture with n species, the forces affecting species i $d_i \cdot \mathrm{R}T \cdot c_{\mathrm{total}}$ can be calculated as the sum of all binary interactions [59]:

$$d_i = \sum_{\substack{j=1 \\ i \neq j}}^{n} d_{i,j} = -\sum_{\substack{j=1 \\ i \neq j}}^{n} \frac{x_i x_j \cdot (v_i - v_j)}{Đ_{i,j}} \tag{A.10}$$

So far, the force which is applied on the molecules of one species i due to their relative velocity to the molecules of other species j has been described. Additionally, due to their gradient in the chemical potential, a drag force acts on the molecules of species i. This drag force is defined for non-ideal systems as $d_i \cdot \mathrm{R}T \cdot c_{\mathrm{total}}$ [60], with [59]:

$$d_i = \frac{x_i}{\mathrm{R}T} \frac{\mathrm{d}\mu_i}{\mathrm{d}z} \tag{A.11}$$

It can be recognized that the drag force of the species i is proportional to its mole fraction x_i and its chemical potential gradient $\frac{\mathrm{d}\mu_i}{\mathrm{d}z}$ [60].
Finally the right side of eq. (A.11) can be substituted in the left side of eq. (A.10):

$$\frac{c_i}{\mathrm{R}T} \frac{\mathrm{d}\mu_i}{\mathrm{d}z} = -\sum_{\substack{j=1 \\ i \neq j}}^{n} \frac{c_j c_i \cdot (v_j - v_i)}{c_{\mathrm{total}} Đ_{i,j}} \tag{A.12}$$

The eq. (A.12) is better known as the (stationary) Maxwell-Stefan diffusion. As illustrated in fig. 3.4, in the Maxwell-Stefan diffusion the drag force of species i is equal to the sum of the interaction forces between the species i and the species j, where $i \neq j$ [59, 60].

Idealization of the Maxwell-Stefan Diffusion

A simplification of the equation (3.20) can be achieved if an ideal behavior of the involved species is assumed. In that case, the concentration can be used as a driving force instead of the chemical potential. The calculation of the chemical potential is given in eq. (A.13), with μ_i^Θ as chemical potential under standard conditions and a_i as the chemical activity of species i [97].

$$\mu_i = \mu_i^\Theta + \mathrm{R}T \cdot \ln(a_i) \tag{A.13}$$

Whereas the chemical activity of species i is calculated by its concentration c_i, its concentration under standard conditions c_i^Θ and its activity coefficient γ_i [97]:

$$a_i = \frac{c_i}{c_i^\Theta} \cdot \gamma_i \tag{A.14}$$

With eq. (A.14), eq. (A.13) can be rewritten as:

$$\mu_i = \mu_i^\Theta + \mathrm{R}T \cdot \ln(c_i) - \mathrm{R}T \cdot \ln(c_i^\Theta) + \mathrm{R}T \cdot \ln(\gamma_i) \tag{A.15}$$

Please remember, the goal of simplification is to replace the gradient in chemical potential by the gradient in concentration as the driving force. Accordingly, the quantity sought is the derivative of the chemical potential over the location. This can be gained by the total differential of eq. (A.15) at a constant temperature:

$$\frac{\partial \mu_i}{\partial z} = \frac{\partial \mu_i^\Theta}{\partial z} + \mathrm{R}T \frac{\partial \ln(c_i)}{\partial z} - \mathrm{R}T \frac{\partial \ln(c_i^\Theta)}{\partial z} + \mathrm{R}T \frac{\partial \ln(\gamma_i)}{\partial z} \tag{A.16}$$

Since μ_i^Θ and $c_i^\Theta = 1\,\mathrm{mol\,L^{-1}}$ are constant and γ_i is defined to be 1 in an ideal system, eq. (A.16) can be reduced to:

$$\frac{\partial \mu_i}{\partial z} = \mathrm{R}T \frac{\partial \ln(c_i)}{\partial z} = \frac{\mathrm{R}T}{c_i} \frac{\partial c_i}{\partial z} \tag{A.17}$$

By substituting eq. (A.17) into eq. (3.17), the flux $\dot{N}_i$ is now expressed in dependence on the gradients in concentration of the species i and j:

$$\dot{N}_i = -\sum_{\substack{j=1 \\ i \neq j}}^{n} Đ_{i,j} \left(\frac{\partial c_i}{\partial z} \frac{c_j}{c_{\text{total}}} - \frac{\partial c_j}{\partial z} \frac{c_i}{c_{\text{total}}} \right) \tag{A.18}$$

With a transformation analog to eq. (3.18) to eq. (3.20) finally the simplified Maxwell-Stefan equation is obtained, which regards the interaction between the species i and j and the gradient in concentration as the driving force. All further before mentioned assumptions are still valid:

$$\frac{\mathrm{d}c_i}{\mathrm{d}t} = \sum_{\substack{j=1 \\ i \neq j}}^{n} Đ_{i,j} \left(\frac{\partial^2 c_i}{\partial z^2} \frac{c_j}{c_{\text{total}}} - \frac{\partial^2 c_j}{\partial z^2} \frac{c_i}{c_{\text{total}}} \right) \tag{A.19}$$

B Specific Values[13]

For all concentration dependent calculations the assumption of a constant total concentration $c_{total} = c_{\text{total}}(z = 0, t = 0)$ is used. As shown in figure B.1 the respective error is less than 0.8 %, i.e.:

$$\left| \frac{c_{\text{total}}(c_{\text{NaOH}}) - c_{\text{total}}(c_{\text{NaOH}}(\text{z} = \text{z}_\text{t}, \text{t}= 0))}{c_{\text{total}}(c_{\text{NaOH}}(\text{z} = \text{z}_\text{t}, \text{t}= 0))} \right| \leq 0.008,$$

$$\text{for } c_{\text{NaOH}} = [1.05 \times 10^4 ... 1.75 \times 10^4 \,\text{mol}\,\text{m}^{-3}] \quad \text{(B.1)}$$

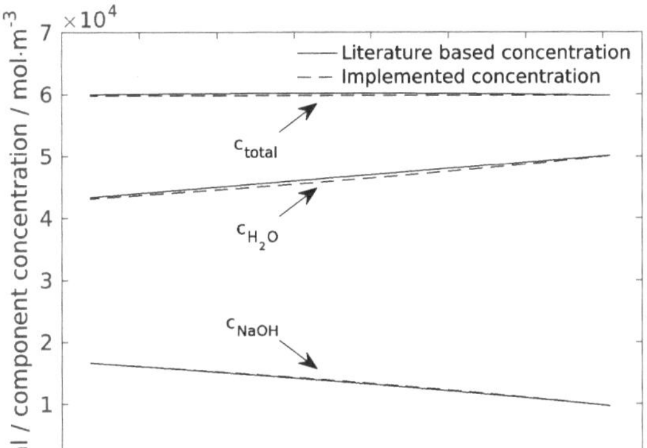

Figure B.1: Total, water, and sodium hydroxide concentration of liquid electrolyte, calculated based on Hirschberg [51] and implemented in the model, depended on the modeled water concentration in the range of the simulation at $T = 353.15\,\text{K}$.

[13] Parts of this chapter have been published in Röhe et al., ChemSusChem, 2019 [1].

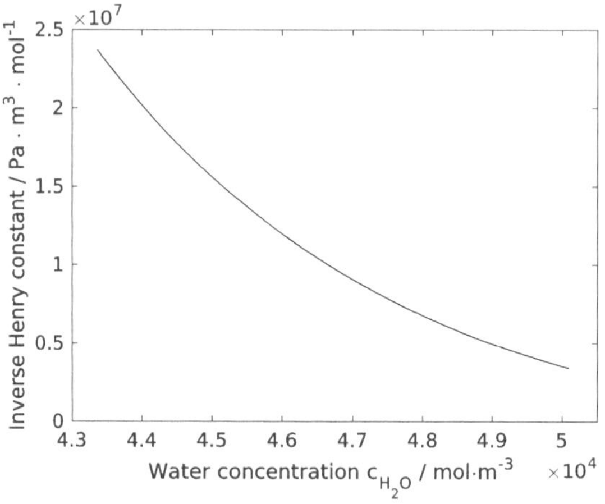

Figure B.2: Inverse Henry constant over water concentration at $T = 353.15\,\mathrm{K}$, hydroxide concentration according to figure B.1.

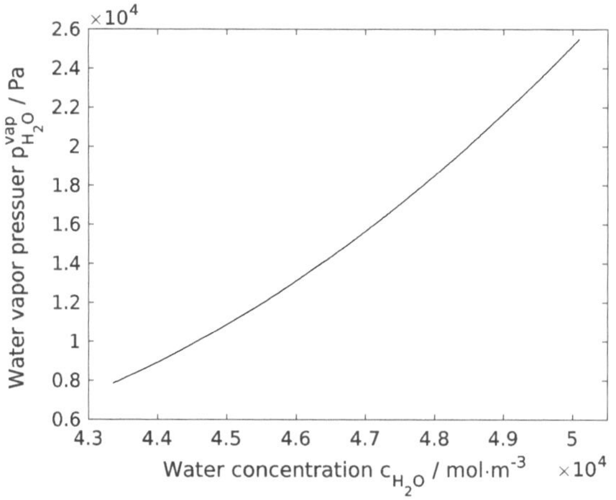

Figure B.3: Vapor pressure of water as a function of water concentration at $T = 353.15\,\mathrm{K}$, hydroxide concentration according to figure B.1.

As seen in figure B.2 the inverse Henry constant is highly sensitive to the water respectively the hydroxide concentration and increases over the simulation range by a factor of 5.81 [22]. Dependent on the water and hydroxide concentration the vapor pressure of water changes decreases over the simulation range by a factor of 2.91 (see fig. B.3) [51].
As shown in figure B.4, based on the enrichment of hydroxide ions the water activity decreases by a factor of 2.22 [75].

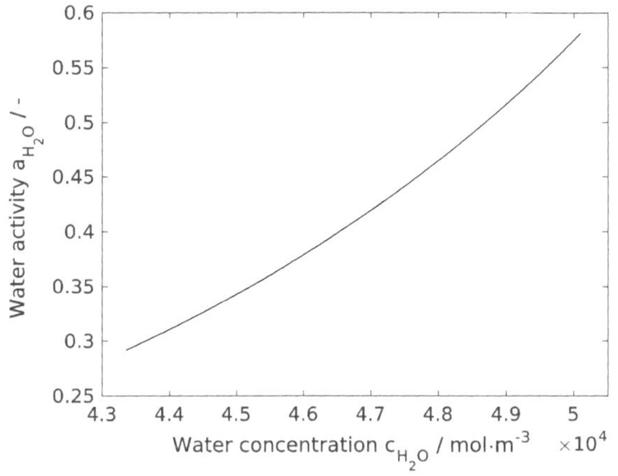

Figure B.4: Activity of water as a function of water concentration at $T = 353.15\,\mathrm{K}$, hydroxide concentration according to figure B.1.

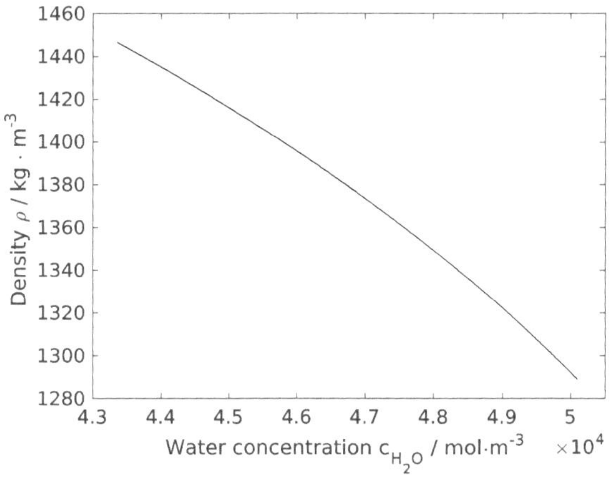

Figure B.5: Density of electrolyte as a function of water concentration at $T = 353.15\,\mathrm{K}$, hydroxide concentration according to figure B.1.

The concentration change of the liquid electrolyte leads to a volume change, and thus to a convective flow (cf. B.5) [68].

C Experimental of Chapter 5[14]

C.1 ODC preparation

The preparation and characterization have been done by Barbara Ellendorff and David Franzen, Clausthal University of Technology, Germany. Section partly written by David Franzen.
The investigated ODCs were fabricated by means of a wet spraying technique [9]. Silver particles (SF9ED Ag Flake, Ames Advanced Materials Corporation, $x_{50} = 1.61\,\mu m$), a methyl cellulose solution (1 wt-%, WALOCEL™ MKX 70000 PP 01) and additional demineralized water were mixed twice with an Ultra Turrax at 13 500 rpm for five minutes until a homogeneous dispersion was obtained. Under a lower rotational speed of 1000 rpm, to prevent agglomeration, a PTFE dispersion (TF 5060GZ, 3M™ Dyneon™) was added. Up to 90 layers of the suspension were applied on a nickel mesh as a conductive supporting material (106 µm x 118 µm mesh size, 63 µm thickness, Haver & Boecker) by using a spraying piston (Evolution, 0.6 mm pin hole, Harder & Steenbeck). The mechanical stability was improved in the following production steps by hot pressing (LaboPress P200S, Vogt, 15 MPa, 130 °C, 5 min) and heat treatment (330 °C, 15 min, air atmosphere), burning out the methylcellulose and sintering the PTFE. This results in a mechanically stable electrode with a pore system that meets the requirements of a gas diffusion electrode. The electrode thickness and the catalyst load were determined using a thickness dial gauge (FD 50, Käfer) and weighing respectively.

C.2 Experimental set-up with electrolyte convection

The measurements have been done by David Franzen, Clausthal University of Technology, Germany. Section partly written by David Franzen.
Electrochemical experiments were carried out in an in-house built half-cell with a flow-through electrolyte and gas chamber separated by the ODC. Measurements were recorded using a Gamry Reference 3000 potentiostat (Gamry Instruments, United Kingdom) with a three-electrode configuration. The ODC was operated as the working electrode, while a platinum wire was placed

[14] Parts of this chapter have been published in Röhe et al., ChemElectroChem, 2019 [2].

downstream in the electrolyte chamber as the counter electrode. Potentials were measured with respect to a reversible hydrogen electrode (RHE) (Hydroflex, Gaskatel, Germany) through a Luggin capillary ending in front of the working electrode. The electrolyte, 30 wt-% NaOH prepared from caustic flakes ($\geq$ 99 wt-%, Carl Roth, Germany) and demineralized water, was pumped through a heating tube and the half-cell from a heated reservoir providing a constant temperature of 50 °C. On the gas side pure oxygen was supplied with a flow rate of $50\,\mathrm{mL_N\,min^{-1}}$. The pressure on the electrolyte and gas side was set to 1.00×10^5 Pa. Sensors on the electrolyte and gas side monitored pressure and temperature. After a startup procedure ($-2\,\mathrm{kA\,m^{-2}}$ hold for 10 min, $-4\,\mathrm{kA\,m^{-2}}$ hold for 50 min, $-2\,\mathrm{kA\,m^{-2}}$ hold for 5 min), potential steps of 100 mV in the range from OCP (1.05 V vs. RHE) to 0.10 V vs. RHE were applied and the current response was recorded.

After 30 s, current density values were extracted to record the polarization curve. For dynamic measurements the potential of 0.8 V vs. RHE was held for 30 s and stepped to 0.7 V vs. RHE. After a further 30 s, the step was reversed in the opposite direction.

C.3 Experimental set-up with SECM and stagnant electrolyte

The measurements have been done by Alexander Botz, Ruhr Universität Bochum, Germany. Section partly written by Alexander Botz.

For SECM the ODC was measured in an in-house built half-cell with a stagnant electrolyte above the electrode and the oxygen flow from below. The whole setup was built on an active damped table (Newport RS 2000, United States) to reduce noise. An in-house built faraday cage was equipped with isolating vacuumed polystyrene panels (Vaku-Isotherm, Germany) to ensure a constant temperature within the time scale of the experiments. The heat-conductive base plate was connected to a thermostat (Huber CC2, Germany) for the direct heating of the electrolyte to the operating temperature 50 °C. The oxygen back pressure ($\geq$ 99.5 %, Air Liquide, France) was regulated via a pressure controller (MKS Instruments, United States). All electronical components and the operator (using a foot connector; Conrad, Germany) were connetcted to the same ground, preventing ground loops and damages at the tip's apex [98].

All electrochemical measurements were conducted utilizing an analogue bi-potentiostat (IPS PG 100, Germany) and the potentials were measured against a commercial RHE reference electrode (Mini Hydroflex, Gaskatel, Germany). A Pt mesh counter electrode was placed behind a thin D5 ceramic frit to prevent platinum species to diffuse to the working electrodes. All NaOH ($\geq$ 98.9 wt-%, J.T. Baker, The Netherlands) solutions were prepared with ultra-pure water (MilliQ, SG, Germany). In the measurements all solutions were air-saturated and heated up to 50 °C in a water bath to the required temperature before being filled into the pre-heated measurement cell.

To ensure stable conditions at the sample electrode a conditioning procedure (cyclic voltammogramms, potential range between 0 to -1 V vs. RHE, 10 cycles, 100 mVs^{-1}) was performed. After ODC conditioning the solution inside the cell was replaced to avoid an activity change of the electrolyte prior to the measurement. During the measurements the backside of the ODC was in contact with oxygen at 1.013×10^5 Pa, allowing passive oxygen transport. First the OCP of the ODC was determined as 1.1 V vs. RHE. Then the ODC was polarized step-wise between 1.1 V and 0.2 V vs. RHE. After the current at the initial potential reached the steady state, the potential was stepped 0.1 V in cathodic direction and the current was recorded for 100 s. From this measurement sequence the polarization and the current dependent NaOH activity were obtained. For dynamic analysis, an additional potential step from 0.8 V to 0.7 V and back to 0.8 V vs. RHE was recorded. Prior to the first step a steady state current was reached. The individual potential steps were recorded for 60 s.

C.4 SECM electrode fabrication and measuring

As tip electrodes for the activity measurements platinum ultra micro electrodes (UME) were employed, which were prepared according to a previously published procedure [99]. Briefly, a laser puller (Sutter Instruments P-2000, United States) was utilized to pull Pt wires (ø: 25 µm, Goodfellow, Germany) inside of quartz glass capillaries ($ø_{out}$: 0.9 mm, $ø_{in}$: 0.3 mm, L: 100 mm, QSIL, Germany) by sealing them in 10 to 12 cycles (20 s laser on and 40 s laser off) with the following parameters: Heat: 800, Filament: 5, Velocity: 128, Delay: 130, Pull: 0. Directly afterwards they were pulled: Heat: 820, Filament: 5, Velocity: 128, Delay: 130, Pull: 160. After soldering the Pt wire with soldering tin (Conrad, Germany) to a copper wire, the electrode tip was polished with an in-house built machine that rotates polishing paper and the electrode individually as described previously [99]. The electrochemical behavior as well as the active electrode size were subsequently determined in a solution of 5 mM $[Ru(NH_3)_6]Cl_3$ ($\geq$ 98 wt-%, Sigma Aldrich, Germany) in 100 mM KCl (100 wt-%, J.T. Baker, The Netherlands).

Pt UMEs were approached to the ODC by means of shear force distance-controlled scanning electrochemical microscopy (SECM). This technique is independent of the current signal and detects hydrodynamic forces in close proximity to the substrate surface [100, 101]. Therefore, two piezo elements (Piezomechanik Pickelmann, Germany) were mounted to the electrode body at an angle of approx. 45°, ca. 1 to 1.5 cm distance from each other and individually connected to a lock-in amplifier (Ametek 7280, Germany). While the upper piezo oscillates the tip apex the lower one detects the resonance movement of the tip. Characteristic resonance frequencies were determined by comparison of frequency spectra (range: 200 to 500 kHz) in the bulk solution and within the shear force interaction distance ($\sim$ 200 nm above the surface). Pre-positioning of the tip at different lateral positions above the ODC surface was controlled using a video microscope

(The ImagingSource USB-camera, Germany). A simplification of the previously reported 4D shear force-based constant-distance mode SECM [102] was used for precise positioning of the SECM tip, utilizing an analogue controlled piezo-based positioning system (PI, Germany). An automatic shear force-based approach curve was performed until a stop criterion of 5 % of the lock-in magnitude value was reached. For all experiments, the tip (ø: ∼ 1 µm) was subsequently retracted to 1 µm to reach the final working distance for the cyclic voltammetry operation. The cyclic voltammetry was recorded at the platinum tip electrode (potential range was chosen according to the expected range between hydrogen and oxygen evolution potentials, $v = 0.05\ Vs^{-1}$) while the ODC was step-wise biased to a potential in between 0 and −1 V against its previously determined OCP. The activity of hydroxide and water was determined as described elsewhere [34]. In a short summary: By detection of cyclic voltammograms at a Pt nanoelectrode in the diffusion field above the ODC surface the Platinum oxide reduction peak was monitored. Considering that the potential of this reaction is dependent on the activity of hydroxide and water, the activity quotient of both species can be extracted applying the Nernst equation. Furthermore, a comparison to a previously recorded calibration curve in different concentrated NaOH solutions enables the direct evaluation of the NaOH concentration.

D Experimental and Scenario Parameters of Chapter 6[15]

D.1 Experimental

D.1.1 Electrode preparation

The preparation and characterization have been done by Barbara Ellendorff and David Franzen, Clausthal University of Technology, Germany. Section partly written by David Franzen.
The preparation method and physical characterization for the examined electrodes is given in detail in a previous publication [23]. In short: For each electrode, a suspension containing silver particles (SF9ED, Ames Advanced Materials Corp.), a methyl cellulose solution (1 wt-% WALOCEL™ MKX 70000 PP 01) as pore building agent and thickener, demineralized water and a PTFE dispersion (TF 5060GZ, 3M™ Dyneon™) in the desired ratio was used. The suspension was supplied with a spraying piston (Evolution, 0.6 mm pin hole, Harder & Steenbeck) on a conductive support (nickel mesh, 106 μm x 118 μm mesh size, 63 μm thickness, Haver & Boecker OHG) with a total of 80 layers. In the further production steps the electrodes were hot pressed (LaboPress P200S, Vogt, 15 MPa, 130 °C, 5 min) and sintered in an air oven (330 °C, 15 min) to improve the mechanical stability and to form the pore system by burning out the methyl cellulose completely. Detailed characteristics of the ODC are also given in a previous publication [23].

D.1.2 Electrochemical characterization

The measurements have been done by David Franzen, Clausthal University of Technology, Germany. Section partly written by David Franzen.
Electrochemical experiments were performed using a half-cell set up (FlexCell HZ PP01, Gaskatel GmbH) with a geometrical cell area of 3.14 cm^2 in a three electrode configuration. A platinum

[15] Parts of this chapter have been published in Röhe et al., Electrochimica acta, 2021 [15].

wire acts as a counter electrode while the potentials were measured with a reversible hydrogen electrode (RHE) through a Luggin capillary in front of the GDE.

The half-cell consists of an electrolyte and gas compartment divided by the GDE. As electrolyte approximately 30 mL of a 30 wt-% NaOH solution was prepared from caustic flakes ($\leq$ 99 wt−%, Carl Roth) and demineralized water. In the gas compartment pure oxygen with a flowrate of 50 mLN min^{-1} and a small back pressure using a 1 mm water column was applied. The whole half-cell was heated up to 80 °C to correspond to industrial conditions [6]. The experiments were carried out with a Zennium Pro Potentiostat (Zahner GmbH) following the same routine for each electrode. In a startup procedure current densities of 1, 2, 3 and 4 $\mathrm{kA\,m^{-2}}$ were applied for conditioning the electrodes. Afterwards pseudo-galvanostatic impedence measurements were performed at current densities of 2 $\mathrm{kA\,m^{-2}}$. A 10 mV amplitude was applied using a single sine with a starting frequency of 1 Hz going up to the upper limit of 50 kHz and sweeping down to the lower limit of 100 mHz. Sampling the frequencies between in the high frequency range twice is a standard proceed to stabilize the measurement. Afterwards linear sweep voltammetry starting at open cell potential (OCP) to 200 mV vs. reversible hydrogen electrode (RHE) with a scan rate of 0.5 $\mathrm{mV\,s^{-1}}$ was performed to obtain the polarization curves.

D.2 EIS at different current densities

I validated the model additional against EIS spectra of the ODC with 98 wt−% Ag (ODC with highest performance) at current densities of 1 $\mathrm{kA\,m^{-2}}$ and 4 $\mathrm{kA\,m^{-2}}$. As presented earlier, the double layer capacitance in ODCs is dependent on the cell potential or the current density respectively [72]. Since this effect is important for the comparison of the impedance spectra at different current densities, the double-layer capacitances for the simulations with 1 and 4 $\mathrm{kA\,m^{-2}}$ have been adjusted according to the in [72] measured relations, all further simulation parameters have not been changed:

- $C_{\mathrm{dl,cat}}(j = 1\,\mathrm{kA\,m^{-2}}) \approx 1.3 \cdot C_{\mathrm{dl,cat}}(j = 2\,\mathrm{kA\,m^{-2}})$
- $C_{\mathrm{dl,cat}}(j = 4\,\mathrm{kA\,m^{-2}}) \approx 0.6 \cdot C_{\mathrm{dl,cat}}(j = 2\,\mathrm{kA\,m^{-2}})$

Measurement and simulation results for different current densities are shown in fig. D.1. The simulation data agree well with the experimental impedance spectra in term of shape, time constant and charge transfer resistance. Both, the measurements as well as the simulations, indicate a acceleration of the time constant with increasing current density. The simulation data show that at all current densities the time constant is dominated by double layer charge and discharge.

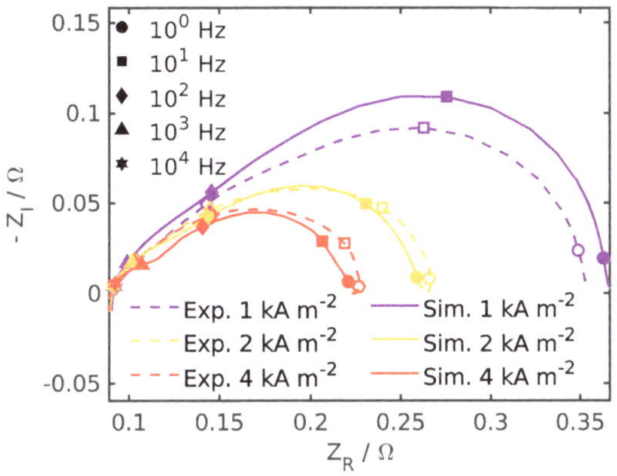

Figure D.1: Simulated and experimental EIS at different current densities of ODC with 98 wt−% Ag.

D.3 Model parameters of the scenario analysis

Please note that the discussed scenarios are hypothetical and based on divergent assumptions. The reaction rate (eq. (6.6)) of the ORR and the charge balance (eq. (6.13)) are functions of the wetted catalyst area. Since the amount of flooded pores and thus the wetted catalyst area are varied in the scenarios, the double layer capacitance and the reaction rate constant have been readjusted to reproduce the measured current densities and time constants.

Semi-homogeneous

- Number of domains: 5
- Specific domain area: Constant for all domains, $S^k_{\text{domain}} = const. = 20\,\%$
- Specific gas-liquid interface: $\frac{S_{\text{phase interface}}}{\Delta z_{\text{3PA}}} = const. = 6.4\,\mu\text{m}^2\,\mu\text{m}^{-3}$
- Δz_{3PA} and z^{k}_{fa} according to table 6.2.

Homogeneous

- Number of domains: 1
- Specific gas-liquid interface: $\frac{S_{\text{phase interface}}}{\Delta z_{\text{3PA}}} = const. = 1.06\,\mu\text{m}^2\,\mu\text{m}^{-3}$, the three-phase area is located directly at the liquid side of the ODC with $\Delta z_{\text{3PA}} = 34\,\mu\text{m}$.

All further parameters are given in table D.1.

Table D.1: Parameters identified from electrochemical experiments and ODC model for ODC with 98 wt−% Ag with hypothetical scenarios.

Name	Symbol	Unit	Semi-homogeneous	Homogeneous
Reaction rate constant	$k_{0,\mathrm{cat}}$	$\mathrm{mol\,m^{-2}\,s^{-1}}$	0.28	0.51
Double layer capacitance	$C_{\mathrm{dl,cat}}$	$\mathrm{F\,m^{-2}}$	3.04	5.21
Porosity[a]	ε	–	0.4	
Specific catalyst interface[a]	S_{cat}	$\mathrm{\mu m^2\,\mu m^{-3}}$	0.79	
Specific resistance[a,b]	R_{specific}	$\Omega\,\mathrm{m^2}$	2.83×10^{-5}	

[a] Same parameter value as identified for the inhomogeneous modeled ODC. [b] High frequency resistance from EIS measurements [23].

Schriftenreihe
des Instituts für Angewandte Materialien

ISSN 2192-9963

Band 1 Prachai Norajitra
Divertor Development for a Future Fusion Power Plant.
ISBN 978-3-86644-738-7

Band 2 Jürgen Prokop
Entwicklung von Spritzgießsonderverfahren zur Herstellung von Mikrobauteilen durch galvanische Replikation.
ISBN 978-3-86644-755-4

Band 3 Theo Fett
New contributions to R-curves and bridging stresses – Applications of weight functions.
ISBN 978-3-86644-836-0

Band 4 Jérôme Acker
Einfluss des Alkali/Niob-Verhältnisses und der Kupferdotierung auf das Sinterverhalten, die Strukturbildung und die Mikrostruktur von bleifreier Piezokeramik $(K_{0,5}Na_{0,5})NbO_3$.
ISBN 978-3-86644-867-4

Band 5 Holger Schwaab
Nichtlineare Modellierung von Ferroelektrika unter Berücksichtigung der elektrischen Leitfähigkeit.
ISBN 978-3-86644-869-8

Band 6 Christian Dethloff
Modeling of Helium Bubble Nucleation and Growth in Neutron Irradiated RAFM Steels.
ISBN 978-3-86644-901-5

Band 7 Jens Reiser
Duktilisierung von Wolfram. Synthese, Analyse und Charakterisierung von Wolframlaminaten aus Wolframfolie.
ISBN 978-3-86644-902-2

Band 8 Andreas Sedlmayr
Experimental Investigations of Deformation Pathways in Nanowires.
ISBN 978-3-86644-905-3

Die Bände sind unter www.ksp.kit.edu als PDF frei verfügbar oder als Druckausgabe bestellbar.

Band 9 Matthias Friedrich Funk
Microstructural stability of nanostructured fcc metals during cyclic deformation and fatigue.
ISBN 978-3-86644-918-3

Band 10 Maximilian Schwenk
Entwicklung und Validierung eines numerischen Simulationsmodells zur Beschreibung der induktiven Ein- und Zweifrequenzrandschichthärtung am Beispiel von vergütetem 42CrMo4.
ISBN 978-3-86644-929-9

Band 11 Matthias Merzkirch
Verformungs- und Schädigungsverhalten der verbundstranggepressten, federstahldrahtverstärkten Aluminiumlegierung EN AW-6082.
ISBN 978-3-86644-933-6

Band 12 Thilo Hammers
Wärmebehandlung und Recken von verbundstranggepressten Luftfahrtprofilen.
ISBN 978-3-86644-947-3

Band 13 Jochen Lohmiller
Investigation of deformation mechanisms in nanocrystalline metals and alloys by in situ synchrotron X-ray diffraction.
ISBN 978-3-86644-962-6

Band 14 Simone Schreijäg
Microstructure and Mechanical Behavior of Deep Drawing DC04 Steel at Different Length Scales.
ISBN 978-3-86644-967-1

Band 15 Zhiming Chen
Modelling the plastic deformation of iron.
ISBN 978-3-86644-968-8

Band 16 Abdullah Fatih Çetinel
Oberflächendefektausheilung und Festigkeitssteigerung von niederdruckspritzgegossenen Mikrobiegebalken aus Zirkoniumdioxid.
ISBN 978-3-86644-976-3

Band 17 Thomas Weber
Entwicklung und Optimierung von gradierten Wolfram/EUROFER97-Verbindungen für Divertorkomponenten.
ISBN 978-3-86644-993-0

Die Bände sind unter www.ksp.kit.edu als PDF frei verfügbar oder als Druckausgabe bestellbar.

Band 18 Melanie Senn
Optimale Prozessführung mit merkmalsbasierter Zustandsverfolgung.
ISBN 978-3-7315-0004-9

Band 19 Christian Mennerich
Phase-field modeling of multi-domain evolution in ferromagnetic shape memory alloys and of polycrystalline thin film growth.
ISBN 978-3-7315-0009-4

Band 20 Spyridon Korres
On-Line Topographic Measurements of Lubricated Metallic Sliding Surfaces.
ISBN 978-3-7315-0017-9

Band 21 Abhik Narayan Choudhury
Quantitative phase-field model for phase transformations in multi-component alloys.
ISBN 978-3-7315-0020-9

Band 22 Oliver Ulrich
Isothermes und thermisch-mechanisches Ermüdungsverhalten von Verbundwerkstoffen mit Durchdringungsgefüge (Preform-MMCs).
ISBN 978-3-7315-0024-7

Band 23 Sofie Burger
High Cycle Fatigue of Al and Cu Thin Films by a Novel High-Throughput Method.
ISBN 978-3-7315-0025-4

Band 24 Michael Teutsch
Entwicklung von elektrochemisch abgeschiedenem LIGA-Ni-Al für Hochtemperatur-MEMS-Anwendungen.
ISBN 978-3-7315-0026-1

Band 25 Wolfgang Rheinheimer
Zur Grenzflächenanisotropie von $SrTiO_3$.
ISBN 978-3-7315-0027-8

Band 26 Ying Chen
Deformation Behavior of Thin Metallic Wires under Tensile and Torsional Loadings.
ISBN 978-3-7315-0049-0

Band 27 Sascha Haller
Gestaltfindung: Untersuchungen zur Kraftkegelmethode.
ISBN 978-3-7315-0050-6

Band 28 Nicht erschienen

Band 29 Gunnar Picht
Einfluss der Korngröße auf ferroelektrische Eigenschaften dotierter $Pb(Zr_{1-x}Ti_x)O_3$ Materialien.
ISBN 978-3-7315-0106-0

Band 30 Esther Held
Eigenspannungsanalyse an Schichtverbunden mittels inkrementeller Bohrlochmethode.
ISBN 978-3-7315-0127-5

Band 31 Pei He
On the structure-property correlation and the evolution of Nanofeatures in 12-13.5% Cr oxide dispersion strengthened ferritic steels.
ISBN 978-3-7315-0141-1

Band 32 Jan Hoffmann
Ferritische ODS-Stähle – Herstellung, Umformung und Strukturanalyse.
ISBN 978-3-7315-0157-2

Band 33 Wiebke Sittel
Entwicklung und Optimierung des Diffusionsschweißens von ODS Legierungen.
ISBN 978-3-7315-0182-4

Band 34 Osama Khalil
Isothermes Kurzzeitermüdungsverhalten der hoch-warmfesten Aluminium-Knetlegierung 2618A (AlCu2Mg1,5Ni).
ISBN 978-3-7315-0208-1

Band 35 Nicht erschienen

Band 36 Christoph Hage
Grundlegende Aspekte des 2K-Metallpulverspritzgießens.
ISBN 978-3-7315-0217-3

Band 37 Bartłomiej Albiński
Instrumentierte Eindringprüfung bei Hochtemperatur für die Charakterisierung bestrahlter Materialien.
ISBN 978-3-7315-0221-0

Band 38 Tim Feser
Untersuchungen zum Einlaufverhalten binärer alpha-Messinglegierungen unter Ölschmierung in Abhängigkeit des Zinkgehaltes.
ISBN 978-3-7315-0224-1

Band 39 Jörg Ettrich
Fluid Flow and Heat Transfer in Cellular Solids.
ISBN 978-3-7315-0241-8

Band 40 Melanie Syha
Microstructure evolution in strontium titanate Investigated by means of grain growth simulations and x-ray diffraction contrast tomography experiments.
ISBN 978-3-7315-0242-5

Band 41 Thomas Haas
Mechanische Zuverlässigkeit von gedruckten und gasförmig abgeschiedenen Schichten auf flexiblem Substrat.
ISBN 978-3-7315-0250-0

Band 42 Aron Kneer
Numerische Untersuchung des Wärmeübertragungsverhaltens in unterschiedlichen porösen Medien.
ISBN 978-3-7315-0252-4

Band 43 Manuel Feuchter
Investigations on Joule heating applications by multiphysical continuum simulations in nanoscale systems.
ISBN 978-3-7315-0261-6

Band 44 Alexander Vondrous
Grain growth behavior and efficient large scale simulations of recrystallization with the phase-field method.
ISBN 978-3-7315-0280-7

Band 45 Tobias Kennerknecht
Fatigue of Micro Molded Materials – Aluminum Bronze and Yttria Stabilized Zirconia.
ISBN 978-3-7315-0293-7

Band 46 Christopher Scherr
Elektrochemisches Verhalten von Lithium-Schwefel-Zellen mit unterschiedlicher Kathodenstruktur.
ISBN 978-3-7315-0296-8

Band 47 Konstantin Frölich
Der Decal-Prozess zur Herstellung katalysatorbeschichteter Membranen für PEM-Brennstoffzellen.
ISBN 978-3-7315-0334-7

Band 48 Benedikt Haspel
Werkstoffanalytische Betrachtung der Eigenschaften von mittels neuartiger RTM-Fertigungsprozesse hergestellten glasfaserverstärkten Polymerverbunden.
ISBN 978-3-7315-0337-8

Band 49 Marco Berghoff
Skalenübergreifende Modellierung und Optimierung vom atomistischen kristallinen Phasenfeldmodell bis zur mesoskopischen Phasenfeldmethode.
ISBN 978-3-7315-0416-0

Band 50 Michael Selzer
Mechanische und Strömungsmechanische Topologieoptimierung mit der Phasenfeldmethode.
ISBN 978-3-7315-0431-3

Band 51 Michael Mahler
Entwicklung einer Auswertemethode für bruchmechanische Versuche an kleinen Proben auf der Basis eines Kohäsivzonenmodells.
ISBN 978-3-7315-0441-2

Band 52 Christoph Bohnert
Numerische Untersuchung des Verformungs- und Bruchverhaltens von einkristallinem Wolfram auf mikroskopischer Ebene.
ISBN 978-3-7315-0444-3

Band 53 Stefan Guth
Schädigung und Lebensdauer von Nickelbasislegierungen unter thermisch-mechanischer Ermüdungsbeanspruchung bei verschiedenen Phasenlagen.
ISBN 978-3-7315-0445-0

Band 54 Markus Klinsmann
The Effects of Internal Stress and Lithium Transport on Fracture in Storage Materials in Lithium-Ion Batteries.
ISBN 978-3-7315-0455-9

Band 55 Thomas Straub
Experimental Investigation of Crack Initiation in Face-Centered Cubic Materials in the High and Very High Cycle Fatigue Regime.
ISBN 978-3-7315-0471-9

Band 56 Maren Lepple
Kupfer- und Eisenoxide als Konversions-Elektrodenmaterialien für Lithium-Ionen-Batterien: Thermodynamische und Elektrochemische Untersuchungen.
ISBN 978-3-7315-0482-5

Band 57 Stefan Andreas Slaby
Charakterisierung und Bewertung der Zug- und Ermüdungseigenschaften von Mikrobauteilen aus 17-4PH Edelstahl. Ein Vergleich von mikropulverspritzgegossenem und konventionell hergestelltem Material.
ISBN 978-3-7315-0484-9

Band 58 Kumar Ankit
Phase-field modeling of microstructural pattern formation in alloys and geological veins.
ISBN 978-3-7315-0491-7

Band 59 Kuo Zhang
Characterization and Modeling of the Ratcheting Behavior of the Ferritic-Martensitic Steel P91.
ISBN 978-3-7315-0503-7

Band 60 Nicht erschienen

Band 61 Fabian Lemke
Untersuchung des Sinterverhaltens von $SrTiO_3$ unter Berücksichtigung der Defektchemie.
ISBN 978-3-7315-0510-5

Band 62 Johannes Kümmel
Detaillierte Analyse der Aufbauschneidenbildung bei der Trockenzerspanung von Stahl C45E mit Berücksichtigung des Werkzeugverschleißes.
ISBN 978-3-7315-0518-1

Band 63 László Hagymási
Modellierung der Stoffübertragung beim Niederdruckcarbonitrieren mit Ammoniak und Acetylen.
ISBN 978-3-7315-0568-6

Band 64 Reza Eslami
A novel micro-mechanical model for prediction of multiaxial high cycle fatigue at small scales.
ISBN 978-3-7315-0583-9

Band 65 Sebastian Schulz
Phase-field simulations of multi-component solidification and coarsening based on thermodynamic datasets.
ISBN 978-3-7315-0618-8

Band 66 Markus Stricker
Die Übertragung von mikrostrukturellen Eigenschaften aus der diskreten Versetzungsdynamik in Kontinuumsbeschreibungen.
ISBN 978-3-7315-0658-4

Band 67 Luis Straßberger
Untersuchung und Modellierung des viskoplastischen Verformungsverhaltens oxidpartikelverstärkter Stähle.
ISBN 978-3-7315-0674-4

Band 68 Mark Wobrock
Microplasticity of idealized single crystalline Ag cantilevers characterized with methods of high resolution.
ISBN 978-3-7315-0682-9

Band 69 Amritesh Kumar
Micromechanical study on the deformation behaviour of directionally solidified NiAl-Cr eutectic composites.
ISBN 978-3-7315-0694-2

Band 70 Johannes Hötzer
Massiv-parallele und großskalige Phasenfeldsimulationen zur Untersuchung der Mikrostrukturentwicklung.
ISBN 978-3-7315-0693-5

Band 71 Thomas Hupfer
Herstellung von LATP für den Einsatz als Festkörperelektrolyt und dessen Eigenschaften.
ISBN 978-3-7315-0702-4

Band 72 Florentin Pottmeyer
Schädigungsverhalten von in CFK-Laminaten eingebetteten Inserts unter bauteilnahen Beanspruchungen.
ISBN 978-3-7315-0719-2

Band 73 Andres Höweling
Untersuchung der Hochvoltstabilität und Tiefentladung von dotierten $LiNi_{0,5}Mn_{1,5}O_4$-Hochvoltspinellen.
ISBN 978-3-7315-0728-4

Band 74 Tabea Gisela Schwark
Deformation and Fracture Properties of the Soft Magnetic Composite Somaloy 700 3P on Different Length Scales.
ISBN 978-3-7315-0759-8

Band 75 Klaudia Lichtenberg
Metallmatrixverbunde mit Verstärkungselementen aus metallischem Glas $Ni_{60}Nb_{20}Ta_{20}$ – Herstellung und Charakterisierung.
ISBN 978-3-7315-0782-6

Band 76 Claudio Findeisen
Charakterisierung und Modellierung von instabilen Metamaterialien.
ISBN 978-3-7315-0869-4

Band 77 Nilesha Mishra
Influence of strain on the functionality of ink-jet printed thin films and devices on flexible substrates.
ISBN 978-3-7315-0853-3

Band 78 Simon Werner Bonk
Plastische Verformungsmechanismen in hochgradig kaltgewalzten, ultrafeinkörnigen Wolframblechen.
ISBN 978-3-7315-0878-6

Band 79 Tim Gräning
Herstellung, Charakterisierung und Optimierung von austenitischen ODS Stählen.
ISBN 978-3-7315-0732-1

Band 80 Peter Rupp
Herstellung, Prüfung und Modellierung neuartiger hybrider Aluminiumschaum-CFK-Sandwichverbunde.
ISBN 978-3-7315-0880-9

Band 81 Benjamin Sebastian Ehreiser
Einfluss mechanischer Lasten auf die Herstellung von Stahl-Glaskeramik-Verbunden.
ISBN 978-3-7315-0954-7

Band 82 Hans Giel
Weiterentwicklung experimenteller Methoden zur Ermittlung thermodynamischer Werkstoffdaten von Lithium-Ionen-Batterien.
ISBN 978-3-7315-0981-3

Band 83 Anna Trauth
Characterisation and Modelling of Continuous-Discontinuous Sheet Moulding Compound Composites for Structural Applications.
ISBN 978-3-7315-0950-9

Band 84 Jonas Johannes Hüther
The Impact of Recycling on the Fibre and the Composite Properties of Carbon Fibre Reinforced Plastics.
ISBN 978-3-7315-0983-7

Band 85 Nicolas A. Mayer
Thermodynamik von Kobaltoxid Anodenmaterialien für Lithium-Ionen-Batterien und ihr elektrochemisches Verhalten.
ISBN 978-3-7315-0996-7

Band 86 Ulrich Führer
Untersuchung und Modellierung des Haltezeiteinflusses auf die zyklische Entfestigung ferritisch-martensitischer Stähle.
ISBN 978-3-7315-0837-3

Band 87 Ebru Cihan
Structure evolution in tribological interfaces studied by multilayer model alloys.
ISBN 978-3-7315-0999-8

Band 88 Markus Sudmanns
Entwicklung einer Kontinuumsbeschreibung für die Versetzungsmobilität in Versetzungsnetzwerken.
ISBN 978-3-7315-1001-7

Band 89 Tao Zhang
Phase-field Modeling of Phase Changes and Mechanical Stresses in Electrode Particles of Secondary Batteries.
ISBN 978-3-7315-1002-4

Band 90 Markus Ganser
On the Electro-Chemo-Mechanical Coupling in Solid State Batteries and its Impact on Morphological Interface Stability.
ISBN 978-3-7315-1047-5

Band 91 Michael Kellner
Modellierung mehrkomponentiger Materialsysteme für die Phasenfeldmethode und Analyse der simulierten Mikrostrukturen.
ISBN 978-3-7315-1044-4

Band 92 Felix Schröckert
Herstellung dünner Folien aus Lithium-Lanthan-Titanat zur Anwendung als Festkörperelektrolyt.
ISBN 978-3-7315-1008-6

Band 93 Ephraim Schoof
Chemomechanische Modellierung der Wärmebehandlung von Stählen mit der Phasenfeldmethode.
ISBN 978-3-7315-1050-5

Band 94 Alexander Valentin Brabänder
Registrierende Härtemessung an neutronenbestrahlten Materialien bei hohen Temperaturen.
ISBN 978-3-7315-1097-0

Band 95 Denny Schmidt
Einfluss der Kompaktierung auf die Elektrodenmikrostruktur und elektrochemische Performance bei Lithium-Ionen-Zellen.
ISBN 978-3-7315-1098-7

Band 96 Svenja Dittrich
Entwicklung von Siebdruckpasten zur Herstellung von Glaslotfügungen für die Festoxidbrennstoffzelle.
ISBN 978-3-7315-1085-7

Band 97 Michael Dippon
Bestimmung der Betriebsgrenzen für das Schnellladen von Lithium-Ionen Batterien.
ISBN 978-3-7315-1123-6

Band 98 Patricia Haremski
Diffusionseigenschaften von Nickel in einer Festoxid-Brennstoffzelle.
ISBN 978-3-7315-1124-3

Die Bände sind unter www.ksp.kit.edu als PDF frei verfügbar oder als Druckausgabe bestellbar.

Band 99 Florian Wankmüller
Mehrskalige Charakterisierung der Hochtemperatur-Brennstoffzelle (SOFC).
ISBN 978-3-7315-1142-7

Band 100 Niklas Russner
Modellgestützte Analyse des Stackbetriebs von Festoxidzellen.
ISBN 978-3-7315-1144-1

Band 101 Theo Fett, Karl Günter Schell, Ethel C. Bucharsky, Gabriele Rizzi, Pascal Hettich, Susanne Wagner, Michael J. Hoffmann
Consequences of hydroxyl generation by the silica/water reaction Part I: Diffusion and Swelling.
ISBN 978-3-7315-1148-9

Band 102 Theo Fett, Karl Günter Schell, Ethel C. Bucharsky, Gabriele Rizzi, Susanne Wagner, Michael J. Hoffmann
Consequences of hydroxyl generation by the silica/water reaction Part II: Global and local Swelling Part III: Damage and Young's Modulus.
ISBN 978-3-7315-1159-5

Band 103 Johannes Dornheim
Modellfreies Lernen optimaler zeitdiskreter Regelungsstrategien für Fertigungsprozesse mit endlichem Zeithorizont.
ISBN 978-3-7315-1158-8

Band 104 Markus Muth
Grundlagenuntersuchungen an intrinsisch gefertigten lasttragenden FVK/Metall-Hybridträgern.
ISBN 978-3-7315-1161-8

Band 105 Oleg Birkholz
Modeling transport properties and electrochemical performance of hierarchically structured lithium-ion battery cathodes using resistor networks and mathematical half-cell models.
ISBN 978-3-7315-1172-4

Band 106 Verena Irene Becker
Modellierung der Mechanik und der effektiven Transporteigenschaften von partikulären Kathoden sowie deren Einfluss auf die elektrochemische Performance von Lithium-Ionen-Batterien.
ISBN 978-3-7315-1174-8

Band 107 Nikolai Zimber
Nanoskalige Analytik der Mikrostruktur von hochdosig bestrahltem Beryllium.
ISBN 978-3-7315-1178-6

Band 108 Francesco Mazzocchi
Development of NbN based Kinetic Inductance Detectors on sapphire and diamond substrates for fusion plasma polarimetric diagnostics.
ISBN 978-3-7315-1181-6

Band 109 Adrian Schmidt
Multiskalige Modellierung von Lithium-Ionen-Batterien.
ISBN 978-3-7315-1227-1

Band 110 Maximilian Röhe
Dynamic Model-based Analysis of Oxygen Reduction Reaction in Gas Diffusion Electrodes.
ISBN 978-3-7315-1234-9